29.50

SYSTEMS METHODS FOR MANAGERS

MANAGERS

A Practical Guide

SYSTEMS METHODS FOR MANAGERS

A Practical Guide

Alan Waring

with contributions from

Bob Clark

and

Tony Boyle

BLACKWELL SCIENTIFIC PUBLICATIONS

OXFORD LONDON EDINBURGH

BOSTON MELBOURNE

Blackwell Scientific Publications
Editorial offices:
Osney Mead, Oxford OX2 0EL
  (*Orders*: Tel: 0865 240201)
8 John Street, London WC1N 2ES
23 Ainslie Place, Edinburgh EH3 6AJ
3 Cambridge Center, Suite 208
  Cambridge, Massachusetts 02142, USA
107 Barry Street, Carlton
  Victoria 3053, Australia

First published 1989

Set by DP Photosetting, Aylesbury, Bucks
Printed and bound in Great Britain by
Redwood Burn Limited,
Trowbridge, Wiltshire,

DISTRIBUTORS

Marston Book Services Ltd
PO Box 87
Oxford OX2 0DT
(*Orders:* Tel: 0865 791155
        Fax: 0865 791927
        Telex: 837515)

USA
  Publishers Business Services
  PO Box 447
  Brookline Village
  Massachusetts 02147
  (*Orders:* Tel: (617) 524 7678)

Canada
  Oxford University Press
  70 Wynford Drive
  Don Mills
  Ontario M3C 1J9
  (*Orders*: Tel: (416) 441-2941)

Australia
  Blackwell Scientific Publications
  (Australia) Pty Ltd
  107 Barry Street
  Carlton, Victoria 3053
  (*Orders*: Tel: (03) 347 0300)

British Library
Cataloguing in Publication Data
Waring, Alan
  Systems methods for managers: a
  practical guide.
  1. Systems
  I. Title   II. Clark, Bob   III. Boyle, Tony
  003

  ISBN 0–632–02606–5

Library of Congress
Cataloging-in-Publication Data
Waring, Alan.
    Systems methods for managers: a
  practical guide/Alan Waring with
  contributions from Bob Clark and
  Tony Boyle.
       p.   cm.
    Includes index.
    ISBN 0–632–02606–5
    1. Problem solving   2. System
    analysis   1. Title.
  HD30.29.W37   1989   89–15689
  658.4'032–dc20   CIP

Contents

Introduction

Whom is this book aimed at?

This book is aimed at a wide readership having a common need, namely how to resolve work-related problems more effectively. At one level, the intended readership encompasses 'managers' in the widest sense who are seeking more than a 'quick fix' in their approach to managerial and/or technical problems. At a somewhat different level, readers will include undergraduates in a wide range of disciplines as well as post-graduates taking, say, an MBA or tackling research projects.

We realise that many readers will know very little about systems whereas others may know quite a bit and some may be relatively experienced. To cater for such different needs, we have structured the content so that beginners can take a leisurely and perhaps limited route if they wish, whereas experienced readers can opt for a 'fast-track' route as outlined in 'How to Use this Book'.

Throughout the book, examples and case studies have been chosen not because they represent problems and issues peculiar to an industry or to particular kinds of reader but precisely because they demonstrate principles common to the readership as a whole. Thus, a case study examining management problems in the brewing industry could quite easily evoke powerful insights into problems of office automation which a particular reader might have. A manager in a factory could learn much about monitoring and control systems from a study of why a serious and costly accident occurred on a construction site.

Practical benefits for you

Unlike many books on systems, this book has the advantage of concentrating on what systems thinking can do for you in practice. If you come to regard this book as a do-it-yourself guide to using systems ideas for practical benefit, we will feel that we have done our job.

You will soon discover that the beauty of systems ideas lies not just in their usefulness in dealing with problems but also in their lack of professional boundaries. By this, we mean that unlike, say, engineering where you really do need to be an engineer in order to design and build a bridge, with systems it does not matter who you are. The book stresses that systems methods are tools for common application rather than 'secret knowledge' to be jealously guarded by particular professional groups. Systems use requires a particular way of *thinking* more than anything else.

Your job function is likely to have some kind of managerial responsibility or potential, even if your job title does not indicate that you manage anything. Throughout the book, we use the term 'analyst' on the assumption that it will be you or someone you hire who will be doing the systems work. So, whether manager, engineer, teacher, vicar, nurse, student or whoever, you should be able to gain some practical benefit from this book by being able to understand problems more clearly, analyse situations more incisively, and devise solutions and action plans that are more defensible than otherwise would be the case.

Our approach to systems

In this book, we use three approaches to systems – 'hard' systems, 'soft' systems, and systems 'failures'. These three approaches have a lot in common as suggested in Figure A yet have distinct features that make each of them ideally suited to particular kinds of use. Most books on systems adopt only the hard approach in some form or other. Soft and failures analyses may be covered only superficially, if at all. We believe that the three aproaches are *complementary*. To use only one method all the time encourages 'tunnel vision' and often produces ineffective results.

We readily acknowledge Professor Checkland's book, the Open University Systems Group and others for inspiring the three-way approach we have adopted. Although this book does not claim to present original systems ideas and techniques, it does present for the first time in a single book *new examples and original case studies of how to use those ideas and techniques*. There are those who will probably shudder at some of the 'impure' systems content of this book. For

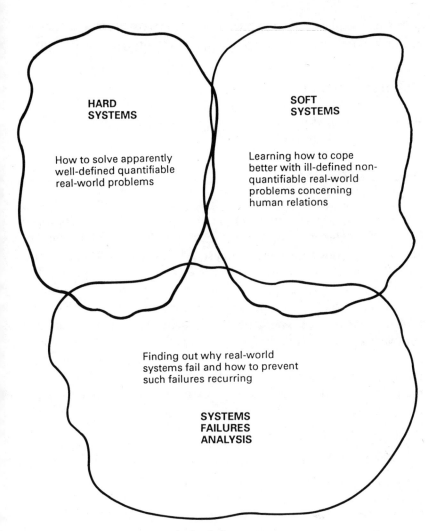

HARD SYSTEMS

How to solve apparently well-defined quantifiable real-world problems

SOFT SYSTEMS

Learning how to cope better with ill-defined non-quantifiable real-world problems concerning human relations

Finding out why real-world systems fail and how to prevent such failures recurring

SYSTEMS FAILURES ANALYSIS

Fig. A. Overview of systems methods for managers.

example, we have used terms such as 'method', 'methodology', 'approach', 'technique' and 'analysis' in a loose, interchangeable way. Wherever possible we have avoided pedantic distinctions because they are unimportant to the reader and the *practical* aim of the book.

Although we cover a broad swathe of systems ideas and techniques, the book has omitted a number of areas. For example, in the operational research area of hard systems we have not covered queueing models. We have not covered planning models such as Gantt charts and critical path analysis. Such omissions were a matter of judgement and we hope that readers will not be disappointed by them.

Organisation of the book

We have organised the book in a way that we think you will find easy to use. There are three parts:

Preview of Part 1

In Part 1 we take an initial look at how to use systems ideas. We have chosen familiar topics from work life to show how a systems approach can improve the effectiveness of tackling problems or uncertainties about how to proceed. We are concerned with 'real-world' problems and not with problems contrived for laboratory examination. The three approaches – hard, soft and failures – are introduced in outline.

Part of the art of systems thinking is making the familiar look strange – being able to look at a perhaps boringly familiar situation in a completely new and stimulating light. For example, imagine you are a farm manager. To a 'non-systems' thinker, a farm is a farm – pleasant, green, smelly, scruffy or whatever. It is where seeds are planted, grow, mature and die. It is where livestock are reared to produce meat, eggs, milk etc. Not a particularly stimulating description. A systems thinker, however, might look at the same farm and describe it like this: 'a habitat system for vegetation and animals (including humans) which is partly controlled by human intervention and from which mutual benefit may be derived.' However pompous this sounds, it does suggest that farms do not just 'happen' and that we as humans are not the only beneficiaries. Try looking at the farm from the point of view of a farm rat or a greenfly! A *systems* view of a farm may be a more fruitful start to planning how to *develop* it, how to *cope* with plant pests, what to *do* when you get bored with the same old crops, or how to *proceed* when a flood of cheap, foreign produce threatens your market position. A systems view may enable you to cope more effectively with labour disputes or to take action to prevent crop failures.

Preview of Part 2

Part 2 takes a closer look at using systems ideas. Systems ideas are very useful tools in tackling the complexities of changes. For example, suppose you were landed with the task of reorganising the work of an office, or a department or even a whole factory. Where would you start? What would you do at 9.00 am on the first day? Panic perhaps? Do your best while relying on experience, intuition and divine providence? Muddle through, telling yourself that it will all work itself out in the end? You may well try to be systematic (a good thing) but that may not be enough. Just look at the many examples of disasters that arise *in spite of* systematic effort! A hard systems approach is *much* more than simply using systematically organised effort.

Let's take another example. Morale in a department has been steadily declining. People mutter in corners; lethargy and lack of enthusiasm abound. Productivity is falling. There is, however, no obvious reason for staff dissatisfaction. Pep talks have not improved things. So, what is wrong? Do you reorganise the staff, perhaps? Liven up the décor? Introduce incentive schemes? Shooting in the dark like this is unlikely to improve things. A soft systems approach, however, may be able to *identify* the underlying reasons for conflict or unease so that remedies may be worked out. A first step in resolving conflicts of interest is to identify the issues and get the interested parties to agree that these *are* the issues.

In some situations, a failure may have occurred. We are all confronted daily with numerous examples, whether in our personal lives, at work, or 'out there'. At the extreme, failures may be described as catastrophes, for example accidents such as King's Cross, Bhopal, Seveso, Flixborough, Chernobyl, Zeebrugge etc. But, small-scale failures may be just as devastating for individuals. A small company going bust can cause great hardship to those who depend on the firm for their livelihood. When things have gone wrong, it is easy to point the finger of blame at 'obvious' causes. However, obvious causes may not be (and rarely are) the underlying causes at all. Preventing similar failures requires understanding of *how* and *why* the failure occurred. Analysing system failures helps to gain that understanding.

In a particular situation, it is often useful to use more than one of the three systems approaches. The final chapter of Part 2 outlines how to pick-and-mix as things develop.

Preview of Part 3

The third part of this book provides a brief academic look at systems thinking for readers wishing to know something about underlying theories. The pick-and-mix theme is also developed to outline how techniques may be woven together to provide a powerful and flexible tool for assisting with the management of change in organisations.

Exercises and activities

All chapters in Part 1 and half the chapters in Part 2 contain exercises. Suggested answers are given at the end of each chapter. Some chapters also contain activities. The purpose of exercises and activities is to get you to do relevant things. An interactive approach is likely to help you learn more effectively than if you simply read through the book.

Realistic expectations

Before diving into Part 1, a word of caution. This introduction may have given you the impression that systems approaches are a kind of philosopher's stone or magic wand to wave at difficult problems or complex situations. Systems thinking is indeed powerful and useful but like anything else it has its limitations; a good craftsman selects his or her tools according to the particular task. We suggest that especially in Part 2 you try to avoid regarding as 'recipes' the system methods described.

Whatever the results of a systems study, there is no guarantee that in every case other people will agree with its findings or suggestions. Even if they express agreement, there may be hidden value systems at work that ensure that changes are not implemented. What systems methods can do is to provide rational tools for analysing problem situations and presenting explicit evaluations that lead to logically defensible decisions.

Acknowledgements

The author would like to thank the following for their support and assistance in the preparation of this book:

Contributors

For contributions to the original concept of the book, for supplying source material in a number of chapters, and for generally contributing to the book's development:

Bob Clark at the Faculty of Technology, The Open University;

Dr Tony Boyle, Managing Director of HASTAM Ltd at Aston Science Park.

Managers

For volunteering to assess and comment on the manuscript as potential users of the book:

Sheila Waring, a Production Manager;

Susan Glendon, a Personnel Administrator with systems analysis experience;

Brian Ward, an Operations Manager with a public utility company;
Gillian Ward, a Company Director.

Management Consultants and Educators

For assessing and commenting on the manuscript during its development and/or providing source material for particular chapters:

Catherine Atthill, formerly Office Technology Unit Project Manager, Southbank Polytechnic;

Professor Richard Booth and Dr Hani Raafat at the Department of Mechanical and Production Engineering, University of Aston;

John Bowden, MBA Course Leader at the London Management Centre, Polytechnic of Central London;

Dr Ian Glendon at the Aston Business School, University of Aston;

Kamran Jalalian, lecturer in computing in further education;

Lynne Jalalian, Department of Environmental Health and Science, Tottenham College of Technology;

John Lewington, a businessman in the United States, formerly at the Department of Management Studies, Harrow College of Higher Education;

Robert Lowe, Senior Consultant with a major venture capital investment bank;

Dr Mike Oatey at the Department of Mathematical Sciences and Computing, South Bank Polytechnic;

Peter Woolfenden, formerly MA Course Leader at the London Management Centre, Polytechnic of Central London.

Other sources of inspiration
Professor Peter Checkland's book *Systems Thinking, Systems Practice*, published by John Wiley & Sons;

Victor Bignell and Joyce Fortune's book *Understanding Systems Failures*, published by Manchester University Press;

Geoff Cutts' book *Structured Systems Analysis and Design Methodology* published by Paradigm Publishing;

Open University course materials on applied systems.

The Publishers

For their unswerving support throughout the book's development:

Jeremy Swinfen Green, Paradigm;

Robin Arnfield, Blackwell;

Dr David Hatter, Series Editor.

How to Use this Book

As noted in the Introduction, there is likely to be a variety of readers having a range of needs. To get the best out of this book, we suggest the following.

Beginners

If you are new to systems or feel that your systems skills are rusty, then work your way through Part 1 at a leisurely pace. You can ignore any Advanced sections in Part 1 chapters. Do as many of the exercises as you feel able. Refer to the Explanation of Terms at the back of the book as necessary.

Read Chapters 1 and 2 first as these provide a general base for the rest of the book. Then read Chapters 3 and 4 together (hard systems ideas), Chapter 5 with Chapter 6 (soft systems ideas) and Chapter 7 with Chapter 8 (systems failures ideas).

Having completed a first pass through Part 1, we suggest you then re-read Part 1 as if you were a 'fast track' reader, this time covering the Advanced sections and doing any exercises and activities that you missed out previously. Having done this, you should then be able to proceed to Part 2 with confidence. Refer to the route map shown in Fig. B.

Fast track readers

As someone with previous systems skills (gained either from Part 1 or from elsewhere) or as someone with a pressing need to apply systems methods, you will not wish to spend too much time on Part 1. We suggest that you start by perusing Part 1 as refresher reading. Read the

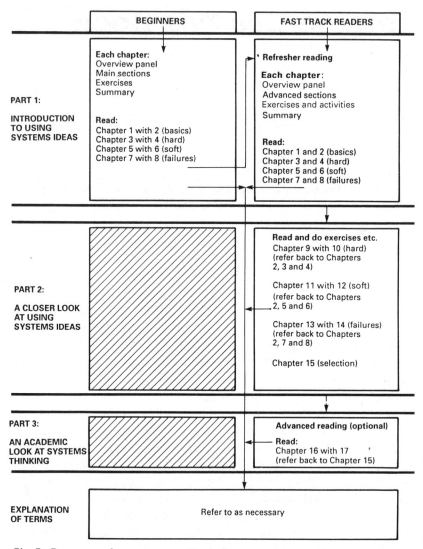

BEGINNERS	FAST TRACK READERS

PART 1:

INTRODUCTION TO USING SYSTEMS IDEAS

BEGINNERS:

Each chapter:
Overview panel
Main sections
Exercises
Summary

Read:
Chapter 1 with 2 (basics)
Chapter 3 with 4 (hard)
Chapter 5 with 6 (soft)
Chapter 7 with 8 (failures)

FAST TRACK READERS:

Refresher reading

Each chapter:
Overview panel
Advanced sections
Exercises and activities
Summary

Read:
Chapter 1 and 2 (basics)
Chapter 3 and 4 (hard)
Chapter 5 and 6 (soft)
Chapter 7 and 8 (failures)

PART 2:

A CLOSER LOOK AT USING SYSTEMS IDEAS

Read and do exercises etc.
Chapter 9 with 10 (hard)
(refer back to Chapters 2, 3 and 4)

Chapter 11 with 12 (soft)
(refer back to Chapters 2, 5 and 6)

Chapter 13 with 14 (failures)
(refer back to Chapters 2, 7 and 8)

Chapter 15 (selection)

PART 3:

AN ACADEMIC LOOK AT SYSTEMS THINKING

Advanced reading (optional)
Read:
Chapter 16 with 17
(refer back to Chapter 15)

EXPLANATION OF TERMS

Refer to as necessary

Fig. B. Route map for systems methods for managers.

Overview Panel at the start of each chapter in Part 1, read any Advanced sections, and make sure that you can do the exercises.

Chapters 1 and 2 cover basic principles. Chapters 3 and 4 introduce hard systems ideas, Chapters 5 and 6 introduce soft systems, and Chapters 7 and 8 cover systems failures.

Having refreshed your knowledge in Part 1, move on to Part 2. Read Chapters 9 and 10 together (hard systems analysis) and refer back to

Chapters 3 and 4 as necessary. Similarly, Chapters 11 and 12 (soft systems analysis) link back to Chapters 5 and 6. The chapters on systems failures analysis (13 and 14) draw on principles developed in Chapters 7 and 8. Finally, read Chapter 15. To get the best out of Part 2, do all the exercises and activities.

Part 3, a short academic section, is optional and is likely to be of interest to those studying systems and organisational behaviour at an advanced level.

PART 1

An Introduction to Using Systems Ideas

Chapter 1
Systems Ideas

Overview

A system as a concept is a recognisable whole that consists of a number of parts that interact in an organised way. Specifically:

- a system does something (there are outputs)
- addition or removal of a component changes the system
- a component is affected by its inclusion in the system
- there are emergent properties, some of which are unpredictable
- a system has a boundary
- outside the boundary is a system environment
- a system is owned by someone.

Systems may be classified as natural, abstract, engineered or human activity systems. Hard systems are mainly natural, abstract and engineered systems whereas soft systems relate to human activity. Hard systems have quantifiable and predictable characteristics whereas soft systems are difficult to quantify or predict, either in how they behave or in their outcomes.

World-view or Weltanschauung refers to a complex set of attitudes, assumptions, motivations, beliefs and values uniquely possessed by an individual or shared by a group of people. World-views will influence how a system behaves and therefore must be identified and examined by the analyst. The analyst must always consider his or her own world-view and how it contributes to an analysis.

1.1 Introduction

The Introduction contains important information about this book, its aims, approach and what you can get out of it. If you have not done so already, please read that Introduction now.

In this first chapter, we are going to introduce some elementary 'systems' principles and get you into the habit of thinking of a system more as an *idea* than as a 'thing'. The word 'system' is used widely in everyday speech and has various meanings. Often it is used to mean something that operates in an organised way, for example a 'systematic procedure' or a 'safe system of work'. However, a systematic procedure is not necessarily a system. A set of logically ordered steps for doing something (a systematic procedure) may have some of, but not all, the necessary ingredients of a system. This kind of popular confusion between system and systematic procedure explains why some so-called systems often fail to meet expectations for their success. How often have you ended up in a mess *despite* using systematic effort?

You have no doubt experienced unintended but nonetheless unwanted effects of other people's work procedures: accounting 'systems', monitoring 'systems', payroll 'systems', stock control 'systems', and so on. The essence of such procedures is that some ingredients or other (inputs) are taken through a procedure (process) and converted into useful products (outputs) as expressed diagrammatically in Fig. 1.1.

This elementary principle of input–process–output is indeed essen-

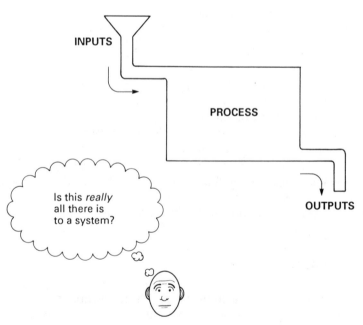

Fig. 1.1 Some people's idea of a system.

tial to all systems but it is rather feeble on its own and encourages a false sense of confidence in those who rely solely on it. There are a lot more things that such an arrangement needs before it properly can be called a *system*.

Advanced panel

We reckon that the kind of confusion just mentioned has its roots in the history of systems work. For several decades, systems practice has been dominated by employer demands for greater efficiency and productivity. The effort of systems analysts, therefore, has been directed towards making the input–process–output procedure more efficient, i.e. it has been ends-and-means dominated. You may be familiar with the idea of feedback by which output is measured and compared against a reference value. If the output is too high or too low, information is fed back to the input side to decrease or increase the input accordingly and so manintain control. Feedback loops and control typify the traditional emphasis on improving the efficiency of a system.

Such traditional systems analysis fits in the area of hard systems covered in later chapters. The most numerous of the traditional systems analysts are those working in the computer and computing field. This numerical and professional domination has had an unfortunate side effect, namely that (a) computing is seen as the only legitimate application for systems thinking, and (b) the methods of computer systems analysts are seen as the only ones of any real value. As mentioned in the main Introduction, we regard systems methods as tools of general application for use by anyone. We do acknowledge the importance of traditional methods, but our approach is to incorporate them within our general 'tool kit' rather than to revere them as being all-powerful.

1.2 What is a system?

As we have said, a system is an idea or concept, even though something that may be called a system will often have a quite obvious physical substance to it. For example, a central heating arrangement for a suite of offices includes pieces of hardware such as hot water radiators, pumps, a boiler, and so on. People often refer to the 'central heating system' meaning all the hardware connected up, but is this really a system? Does it possess all the necessary ingredients? At a very simple level it may be called a system, but the arrangement as described above is not a system in our terms.

At a simple level, a system is: a recognisable whole that consists of a number of parts (called components or elements) that are connected up in an organised way (the system's structure); the components interact, i.e. there are processes going on.

This baseline definition would cover the popular idea of interconnected parts and processes as in the central heating system. However, to be of real use such a definition needs amplifying as is done in the following section.

1.3 Structures and processes

The system's structure is represented by relatively stable, lasting components, i.e. the 'doers' and the 'done to'. The processes within the system are represented by transient, changing components, i.e. actions, change, growth, decline, 'doing'.

However, a system not only has interconnected parts that interact, it also has the following characteristics:

(1) a system does something (there are outputs)
(2) addition or removal of a component changes the system
(3) inclusion of a component affects the component
(4) a system has 'emergent properties'
(5) a system has a boundary
(6) a system has an environment (outside the boundary) that affects it
(7) someone 'owns' the system (i.e. is interested in it for the purposes of study, improvement etc.).

Exercises

1. Taking the office central heating system as an example, (a) identify the outputs, (b) predict the likely effects of removing a room thermostat (temperature controller), and (c) say how the system's environment might affect this system.

The answer to the previous exercise may have surprised you. When people think of system output, they usually do so in terms of the expected or desired output. Nevertheless, all systems have multiple outputs, some of which may be undesirable.

Just as addition of a component changes the system, so too is the component itself affected. For example, the central heating pump receives electrical power and should respond to control signals from the thermostat. The pump will get warm during use and undergo wear and tear. To widen your thinking on this, pause for a moment and consider what systems *you* are a component of and how your presence or absence from them affects them and you.

Three of the characteristics listed above (environment, emergence and ownership) require further explanation as follows.

System environment

The concept of a system environment is important. The environment comprises components that affect the system but which the system is unable to control and is unlikely to affect. The central heating example is *designed* to affect its physical environment (i.e. the air temperature) and so prevent thermal discomfort among the occupants of rooms served. However, the system does not exercise control over its environment. Figure 1.2 shows the system (within the boundary) and people in offices forming part of the environment. Figure 1.2a depicts the physical components of the system and Fig. 1.2b a slightly more 'systemsy' view.

To use a different kind of system as another example, the environment of a 'bridge building' system might include:

- financial restrictions
- legislation
- local authority planning committee
- Department of the Environment
- weather
- geological conditions.

These components affect the system rather than the other way round.

Exercises

2. What other components might be present in the environment of a central heating system?

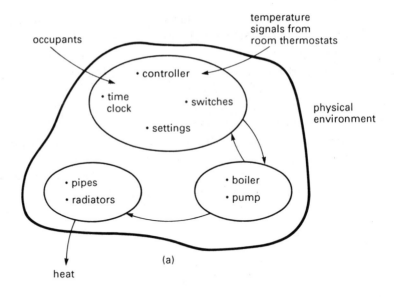

(a)

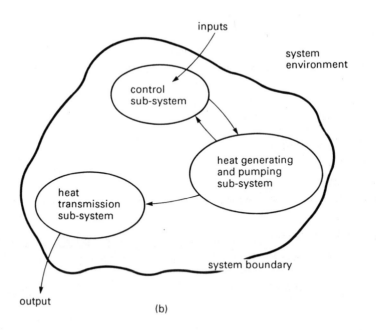

(b)

Fig. 1.2 Office central heating system.

By now, you should be realising that a system encompasses many more components than the more obvious ones. But, who is to say which components are 'correct', and how will you know when you have included all the relevant ones? There are no easy answers to these questions. Systems include necessary components (i.e. those that are essential to any system of its type) and additional components which vary according to the situation and the view of the analyst. Clearly, the hot water central heating system would not function at all without a boiler and a pump. These are essential components. Whether or not to include a thermostat in the system depends on the analyst's viewpoint. As we have seen, there are other important components that would depend on the particular situation. Knowing when to say 'enough is enough' is a matter of experience, systems skills and taking account of your purpose. The process of 'iteration' (or repeated refinement until you decide enough is enough) is central to the methods in this book and is described more fully in Chapter 2.

Emergent properties

Emergence is another property important to the concept of a system. A system is not just a collection of interconnected components. Their interaction as a whole or 'synergy' produces emergent properties or behaviour that could not readily be predicted simply by examining each component in isolation or even some of the components together. For example, examination of a microchip does not enable you to predict all the functions of a complete microcomputer system. Examination of a company's production department (a sub-system of the formally organised system of the company) does not enable prediction of the company's corporate image in the eyes of its customers. Emergence is consistent with holism or, in popular speech, 'the whole is greater than the sum of its parts'. Scrutiny of individual components and sub-systems may provide some useful information about the system as a whole, but it will be limited.

Systems have a mixture of predictable (usually desirable) properties and properties that are difficult to predict. These less obvious properties are often overlooked when systems are being designed. Perhaps we should be more precise. Many undesirable yet predictable properties are often overlooked; there will always be some truly unpredictable properties. The job of systems specifiers and designers is to look beyond the expected, desirable outcomes and try to identify undesirable possibilities so that they may be designed out.

Prediction of adverse properties is especially important where safety is concerned. For example, power presses (powered metal pressing systems) have a habit of operating unexpectedly and numerous cases of power press operators having fingers amputated are on record. Some components can often work together counter-intuitively or do the opposite to what might be expected. Systems designed with noble intentions (e.g. social aid schemes, housing corporations etc.) sometimes can have catastrophic consequences for those they were meant to help. Examples of such system failures are given in Chapters 7, 8, 13 and 14.

Who 'owns' the system?

The idea of system owner may seem a little bizarre. Surely it is clear who the owner is or who the owners are? Not necessarily. You as a manager may exercise control over many things but when one of them becomes problematic you may require that system to be studied in some depth, either by you personally or by others. In this context, you rather than your employer 'own' the system.

The concept of ownership is important because it relates to who controls and maintains the system. Your employer may own the property and equipment and employ the people involved, but you are the one with the 'naughty' system that is causing you sleepless nights! If you delegate the problem solving to your assistant and he or she calls in a consultant, then your assistant becomes the consultant's 'client'; you remain the system/problem owner. These terms 'owner' and 'client' and their implications are discussed more fully, particularly in Chapters 3, 4, 9 and 10.

1.4 Types of system

For the purposes of this book, systems may be classified under four main headings:

- natural systems (e.g. biological systems)
- abstract systems (e.g. computer programs, simulatory models)
- engineered systems (e.g. computer hardware, automobiles)
- human activity systems (e.g. a company, a department, a committee).

This classification enables some kind of sense to be made of the enormous number and range of things that can be described as systems. Although all four types share common features, each also has its own characteristics.

Exercises

3. Allocate each of the following to the most appropriate system type (natural, abstract, engineered, or human activity):

   nuclear reactor
   Burmese teak forest
   strip steel mill
   company marketing department
   taxation
   AIDS epidemic
   warehouse
   central heating system
   housing association.

Systems can also be classified as either 'hard' or 'soft'. Engineered systems, natural systems, and many abstract systems are 'hard'. Hard systems typically have (or are assumed to have) a high degree of predictability vis-à-vis their properties. Hard systems are characterised by having readily *quantifiable and measurable attributes*. Compared with soft systems, hard systems have fewer unpredictable properties but their possibility should always be borne in mind.

Soft systems have a much higher degree of unpredictability because they involve (or are perceived to involve) *people's attitudes and behaviour* which are difficult to predict. Soft systems typically have properties that are difficult to quantify and measure, e.g. viewpoints, conflicts, vested interests and other qualitative aspects.

Exercises

4. Allocate the following to either the hard or soft system category:

   (a) lorry (motor vehicle system for transporting heavy goods?)
   (b) computer steering committee (a system for guiding the introduction and implementation of computers?)

    (c) parent–teacher association (a system for enabling parents and teachers to cooperate to improve children's education?)
    (d) a trout farm (a system for rearing trout for profit?).

1.5 World-view

World-view or Weltanschauung (pronounced velt-an-sha-ung) meaning world-view is a term that you will meet throughout this book. From now on, we will refer to world-view and Weltanschauung interchangeably and use W/a for short.

World-view represents the complex set of perceptions, attitudes, beliefs, values and motivations that characterise an individual or group. It is a 'soft', mental phenomenon that cannot be measured as such, only inferred from what people say or do. Its importance in systems work is immense. Not only must you consider the world-views of 'actors' in the setting you are investigating but you must also be explicit about your own world-view. World-view is a kind of perceptual 'window' or 'tinted spectacles' through which each of us interprets the world and him or herself. Your own world-view is your unique set of biases that, for example, makes you tend to judge situations and people in a characteristic way as in 'value judgements'.

Advanced panel

The major implication of world-view is that you cannot safely assume that everyone's interpretation of a system, or indeed anything, is the same. It should not be taken for granted that everyone where you work perceives your job function the way you do or shares your views about problems, issues, solutions and priorities. This may seem obvious when stated like this yet the fact that W/as are so often *not* shared is frequently overlooked. In systems work, it would be impossible to consider the world-view of each individual in, say, an organisation employing 50, 100 or maybe more people. However, it is important to identify world-views of key individuals who appear to exert particular influence.

Of course, many people do have *similar* world-views. Such collective world-views both drive and are driven by corporate cultures, i.e. sets of values held in common by various groups in an organisation. This is important for the integration and stability of the organisation. However, sometimes a shared view becomes so embedded that it obscures unpalatable facts that tend to contradict that view. The most obvious examples are the ideological stances of political parties. In work organisations, stubborn or dogmatic adherence to a particular world-view can be damaging. For example, for a long time the role of the medical professions was regarded as one of intervention to cure disease, e.g. treatment by surgery or by drugs. Hospitals were run on this assumption. Nowadays, much more emphasis is placed on preventing ill-health and on assessing the whole patient (holistic medicine) in order to establish causes rather than to provide treatment for signs and symptoms. As another example, some magazine publishers are dominated by the world-view of their advertising staff to an extent that editorial content suffers and readership stagnates or declines. A case of this is covered in Chapters 5 and 6.

Exercises

5. World-views have to be considered in analysing hard systems, soft systems and systems failures. Why? Does any one of the three systems approaches warrant a deeper examination of world-views? If yes, which one and why?

C. West Churchman (1985) wrote a very amusing and illuminating piece in which he relates a fable about an aircraft on route to New York being hijacked to Cuba. On board as the plane heads for Cuba, three passengers pass the time in discussion about the nature of reality and how to improve things. They are a hard systems expert, a high powered executive and a professor of philosophy. Each argues that the approach of their profession to social problems is the best. A fourth passenger sitting with them gets steadily drunk and contributes nothing to the debate other than swearing occasionally.

The debate gets very heated with none of the three giving ground to the others. Each is convinced that the world-view of his profession is superior and enables problems to be solved much more effectively. After several hours of argument, the pilot announces that the hijackers have ordered him to fly to New York to release the passengers. The three debaters demand to know why.

At this point, the fourth and rather drunk passenger responds.

'Because I'm one of the hijackers. At first we figured to rob the USA of some of its high-powered talent. But after listening to the lot of you, it's obvious you'll do harm wherever you are with your constant talk, talk, talk! So we're taking you home. Revolution is the only way the oppressed people can win, and revolution will win while you're all busy debating world-views.'

As the hijacker disappears, the stewardess apologises for his drunken outburst and then adds: 'But then he had a point, you know. After all, none of you men ever once included a woman in your so elegant and comprehensive systems.'

Activity

As Churchman's fable shows, the world-view of a particular professional or other group creates a tendency towards a biased view of issues, problems and solutions. We recommend that you spend some time thinking about your own membership of groups and your world-view. How does your world-view affect your capacity to 'put yourself in other people's shoes'? (i.e. to imagine Weltanschauungen that may differ radically from your own). Try to discard any notion you may have that your W/a is 'good/correct' and ones that you do not share are 'bad/incorrect' as it will interfere with your development of systems skills.

Sometimes world-views differ so much that either a clash occurs or people part company. In boardrooms, blood is sometimes said to 'gush under the doors'. In industry, managements and unions fight or manoeuvre to secure advantage in the play for valued resources. The battles between the trade unions and employers provide classic examples of views that are often 'worlds apart'.

On a grand scale, differences in political and religious ideologies result in conflicts and wars between peoples and countries, the Middle East, Afghanistan and Central America providing contemporary examples. Albania has effectively shut out the rest of the world for the past 40 years.

In all systems work, the analyst has to try to identify implicit and explicit world-views in the situation. In analysing soft systems, W/as are especially important since they are usually at the heart of the 'messy' problems being experienced.

1.6 Summary

Systems do not exist in an absolute sense; a system is an idea, a convenient metaphor for a wide range of things. In order to warrant being called a system, it must have:

- a structure of interconnected components
- processes
- a boundary
- an environment comprising components outside the boundary that tend to affect the system rather than be affected by it
- emergent properties
- an owner.

System types may be categorised as 'hard' (natural, abstract, engineered) systems or 'soft' (human activity) systems. The world-views of key individuals and groups who are in or who affect a system are very important in understanding system behaviour which must be gained before attempting to change it.

1.7 Suggested answers to exercises

1. (a) The main output is convective heat from the so-called 'radiators'. The radiator warms the air which rises and moves around the room (less than 25% of heat output is actually radiated in the scientific sense). Other outputs include sound (e.g. pump noise, vibrations, air locks) and waste gases from the boiler flue.
   (b) A room thermostat contains a device designed to monitor air temperature in the room and to switch the boiler on or off. If the temperature is higher than a set value, the boiler is switched off; if lower, the boiler is switched on. Removing the thermostat would make it difficult for the air temperature to be controlled; removal of a component *changes* the system.
   (c) The air temperature itself would affect the system provided that a thermostat was operating. So also would the presence or absence of people in the environment, whether in the offices or in the boiler room. People may open or close windows. They may switch on or off local control valves on radiators. Building services engineers and maintenance staff may adjust the equipment.

2. Other possible components in the environment of such a central heating system:

- electricity supplies
- gas supplies
- boiler operatives
- maintenance engineers
- statutory boiler inspectors
- insurance surveyors
- legislation
- company budgets.

Most of these components affect the system but are affected by it very little.

3. *Natural systems:* Burmese teak forest, AIDS epidemic.
   Abstract system: taxation.
   Engineered systems: nuclear reactor, strip steel mill, warehouse*, central heating system.
   Human activity systems: company marketing department, housing association.

   [**Comment:* We thought of 'warehouse' as an engineered system of equipment (racking, fork lift trucks, conveyors etc.) and processes (goods inwards, outwards, record keeping etc.). However, it is possible to argue that a warehouse is a human activity system if you think of it as 'warehousing'. The term 'socio-technical system' is sometimes used for systems that incorporate both technical and human activity. How you *name* your system is thus important – see the questions in Exercise 4, for example.]

4. *Hard systems:* (a) lorry, (d) trout farm.
   Soft systems: (b) computer steering committee, (c) parent–teacher association.

   (b) and (c) are essentially soft because they require political interaction and debate in order to function.
   (a) is obviously an engineered system of hardware.
   (d) is explicitly a biological system manipulated and controlled by man; it could become soft if, for example, relations between management and staff of the trout farm deteriorated. Remember

that you as observer decide whether to view a particular system as hard or soft.

5. A system does not exist other than in the minds of people. The world-views of those people who are associated in some way with a particular system will influence how that system behaves – what it does, how it does it and whether it avoids failure. The world-view of the analyst will also be relevant. This holds true for hard systems, soft systems and systems failures.

Soft systems require a deeper examination of W/a because world-views of particular actors or groups in the human activity system are likely to be at the heart of that system's malaise.

Chapter 2
Developing Systems Thinking

Overview

The three main types of diagram are (a) structure and relationship (organisation charts, system maps, and influence diagrams), (b) process diagrams (flow charts, decision sequence, flow block, and data flow), and (c) thinking aids (rich pictures and spray diagrams). The main conventions or diagramming rules are Venn and digraph. To avoid confusion, different types and conventions should not be used by unskilled analysts in one diagram.

Systems description provides an invaluable way in to a situation before committing yourself to a particular full-blown analysis. Systems description as a pre-analysis consists of awareness (spray diagrams, rich pictures), commitments (of you and client), tests (for systems), separation and trial boundaries, and selection of key systems.

2.1 Introduction

In Chapter 1, we introduced you to the concept of a system. In this chapter, we aim to develop your systems thinking and in particular introduce you to techniques for describing and visualising systems. This chapter is fairly meaty and has a lot of exercises and activities. We suggest therefore that you take a fairly leisurely approach to it.

2.2 Diagramming techniques

Diagramming techniques are essential to good systems practice because they:

- require the analyst to think clearly about the topic of interest; often they cause the analyst to revise his or her thoughts on the topic

- form a permanent record of the analyst's thoughts on the subject at the time, and are useful for future reference
- enable easier communication of information between analyst and others.

We offer a word of comfort to readers who are not good at drawing and who are already paling at the thought of drawing system diagrams. Artistic talent is definitely not required, and sketching and scribbling skills will be adequate for most needs. Only if and when you have to present diagrams in reports or to meetings will you need to tidy them up.

There are many diagramming techniques used in systems work and these are summarised in Fig. 2.1. This figure represents diagramming techniques in the form of an organisation chart and so is itself an example of a system diagramming technique (i.e. a structure and relationship diagram).

Some techniques are usually associated with particular kinds of work, but these industry-specific techniques such as data flow diagrams and entity-relationship diagrams as used in information systems development are variations and adaptations of the more general techniques shown in Fig. 2.1. If you are especially interested in

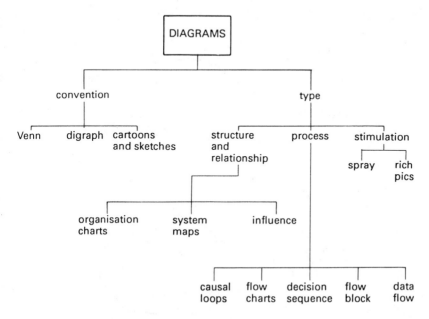

Fig. 2.1 Organisation chart of diagramming techniques.

information systems, we recommend that after reading our book you read *Business Information Systems* by Chris Clare and Peri Loucopoulos in the Paradigm Computer Studies Series.

Organisation charts

Activity

Sketch an organisation chart for an organisation that you know e.g. your department, your employer, a club, your family.

Organisation charts introduce two concepts that are important in all systems work. First is the concept of relationship. For example, Fig. 2.1 indicates that flow block diagrams are a type of process diagram;

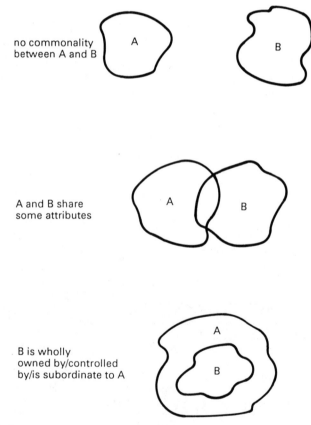

Fig. 2.2 Conventions in Venn diagramming.

flow block diagrams relate to process rather than to structure and relationship. As another example from Fig. 2.1, Venn diagrams relate to conventions or rules for drawing diagrams rather than to types of diagram.

The second concept is that of hierarchies in relationships – the idea that one item is subordinate or junior to another or that one item is a more specific example of another. Very often you will be in possession of a jumble of information. By drawing up a list of topics and then putting like with like, you can readily construct an organisation chart (or other structure and relationship diagram) that summarises and clarifies that information.

Exercises

1. Look at Fig. 2.1 again. Assuming that there is nothing missing from the organisation chart itself, is there anything else that such a diagram ought to have on it?

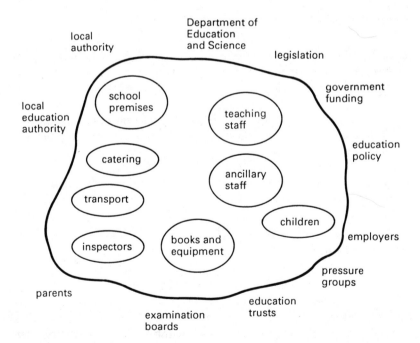

Fig. 2.3 System map of a school.

2. We have just stated that an organisation chart helps to summarise and clarify relationships. This can be very useful in analysing a business organisation, for example. Can you foresee a downside to such use?

System maps

A system map is often drawn according to the Venn convention. This convention is consistent with the theory of 'sets' whereby sets of like components are bounded together and boundaries overlap where some properties of two or more sets are shared as shown in Fig. 2.2.

The following examples show how the Venn convention has been used to construct system maps. Fig. 2.3 is a simple map of a school and Fig. 2.4, which concerns a typical local authority housing department, may look daunting but bear in mind that it started out as a simple rough-and-ready diagram like Fig. 2.3. You do not need to make the components in your system maps look as regular as they do in Fig. 2.4 – using straight lines rather than blobs is a matter of preference.

Exercises

3. In Fig. 2.4, how would you interpret the relationship between the assessments function of the housing department and the customers?

4. How would you interpret the relationship between the department's system for building new houses ('new build') and external suppliers of services?

5. How would you interpret the relationship between HM Government and the housing department?

Advanced panel

Figure 2.4 also introduces the concept of a wider system. A housing department does not enjoy an independent existence; it operates entirely within and under the control of the local authority. The limits of the local authority itself are shown as the wider system boundary outside which the system environment affects both the wider system

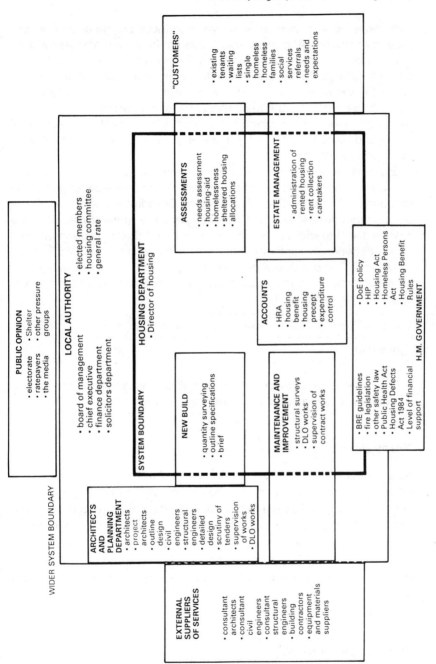

Fig. 2.4 System map of a typical local authority housing department. (*Note:* purpose is to clarify relationships between components and adjust resolution to a manageable level.)

and the system of interest. If we were to focus on the accounts sub-system of the housing department, this could be regarded as existing within a wider system represented by the housing department. Expenditure control could be seen as a sub-system of accounts if *accounts* is the main focus or as a system within the wider accounts system if *expenditure control* is the main focus. Thus, any particular system may be regarded as part of a hierarchy or nest of systems and the setting of boundaries (sub-system, system, wider system) is determined by your focus of interest.

Systems work requires the analyst to make a judgement as to what level he or she is working at, i.e. the *resolution* or level of detail. During iterations, the resolution often has to be adjusted so that all the components (areas) are at the same level. The general rule is that a system diagram should not contain more than a dozen components and ideally between five and nine. Figure 2.4 is therefore stretching this rule to the limit. The practical significance of this rule is that it allows you to adjust the work to manageable levels in terms of conceptual complexity and your resources available.

Exercises

6. Imagine that you work for a company that runs management training courses. Draw a simple system map of your market.

Spray diagrams

Spray diagrams (or 'spider grams') are useful for loosening up your thinking in the early stages of problem solving or analysis. Typically, in the early stages the analyst's thinking is dogged by partial images of 'the problem' and of possible solutions based on past experience. Systems thinking tries to break out of this serial, 'vertical' approach by forcing an expansion of thought i.e. 'lateral thinking'. Alternate expansion and contraction of focus is a common feature of systems work.

Free car parks to go!

A minimum fee of 20p per car is soon to be levied on motorists who presently use six free car parks in Blogton, according to the local council.

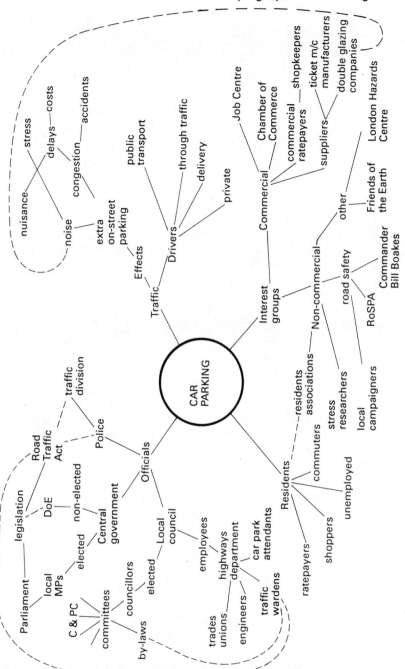

Fig. 2.5 Spray diagram of components relating to car parking situation (first iteration).

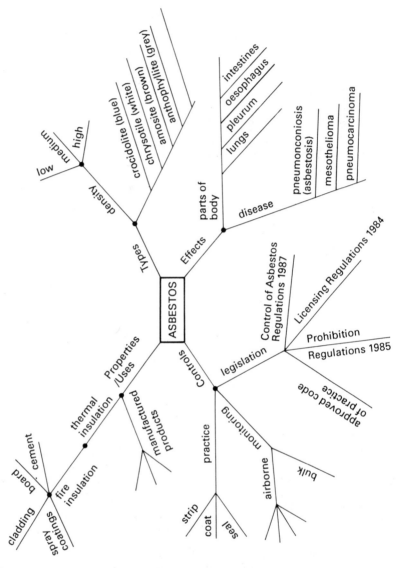

Fig. 2.6 Spray diagram of the asbestos situation.

The council's Coordination and Policy Committee approved the scheme last week and it is to be considered again by the full council next Wednesday.

A one-month experiment will be conducted initially to see what level of takings results. The six car parks have already been assessed with regard to use, size and likely transfer to on-street parking if the charges are introduced. Experimental charging will cost initially £600 per site plus wages of about £120 per week per site. If successful, the experiment will lead to 'pay-and-display' ticket machines. Local shop-keepers and union officials representing traffic wardens are thought to be against the new proposals. (*The Blogton Advertiser*, 1984)

The reasons for the car park levy proposal are not stated but what could the effects be? Who could be affected? The knee-jerk response of vertical thinking is to focus on the obvious effects – car drivers having to pay or park elsewhere. But, introduction of a levy represents a new component in the road traffic system of Blogton and, as you know, a new component will *change* the system. Will the new system behave differently?

Figure 2.5 shows our spray diagram that opened up our thoughts on the subject. We pursued particular lines as indicated but you might have chosen other lines of enquiry. It is also possible to draw spray diagrams in a more structured way as in Fig. 2.6 which relates to asbestos.

Activity

Draw a spray diagram of a subject that is causing you problems or is otherwise important to you.

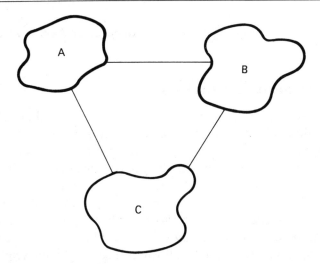

Fig. 2.7 Simple influence diagram (digraph convention).

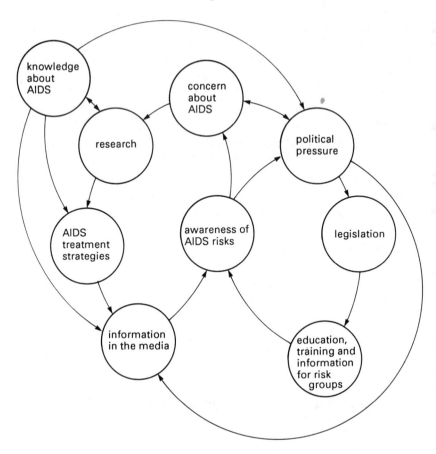

Fig. 2.8 Influence diagram relating to awareness of AIDS risks (digraph convention).

Influence and causal diagrams

The digraph convention employs linked boxes or blobs although many analysts do not bother to enclose components in circles or 'blobs'. At the simplest level, the lines between components only show that a relationship exists but indicate nothing about the direction of influence as in Fig. 2.7. Arrowheads can be used to indicate direction of influence as in Fig. 2.8. Double-headed arrows indicate mutual influence.

Influence means that one component affects another but that does not mean necessarily that a *causal* relationship exists between them. For example, a person's height influences their weight but weight is not caused by height. Causal diagrams are used where evidence suggests a causal process. For example, there is evidence to suggest that an increase in accident prevention activity causes a reduction in the number of accidents (see Fig. 2.9a). In Fig. 2.9a, the minus sign at the arrowhead indicates a reduction. However, there is also evidence to suggest that as the number of accidents increase so does the amount of accident prevention activity (Fig. 2.9b). The plus sign at the arrowhead in Fig. 2.9b indicates an increase. Now, we can combine these two causal diagrams into a single *causal loop diagram* (Fig. 2.9c).

Figure 2.9c is read thus: 'as the number of accidents increase, the amount of accident prevention activity increases; as the accident prevention activity increases, the number of accidents decrease.' The

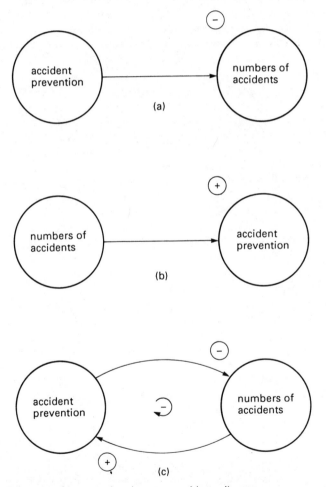

Fig. 2.9 Example of how to develop a causal loop diagram.

minus sign in the middle of the loop indicates control and the arrow shows the direction. The net effect of this causal loop is the tendency to limit, or control, the number of accidents (i.e. a negative feedback loop). Figure 2.10 shows an expanded version with several causal loops. Thicker lines indicate what the analyst considers to be the main causal links. A plus sign in the middle of a loop indicates that it is self-enhancing or self-maintaining (i.e. overall growth or overall decline). A positive loop contains either none or an even number of minus arrows. A negative loop contains an odd number of minus arrows. Figure 2.10 is in effect a multiple cause diagram that incorporates several causal loops.

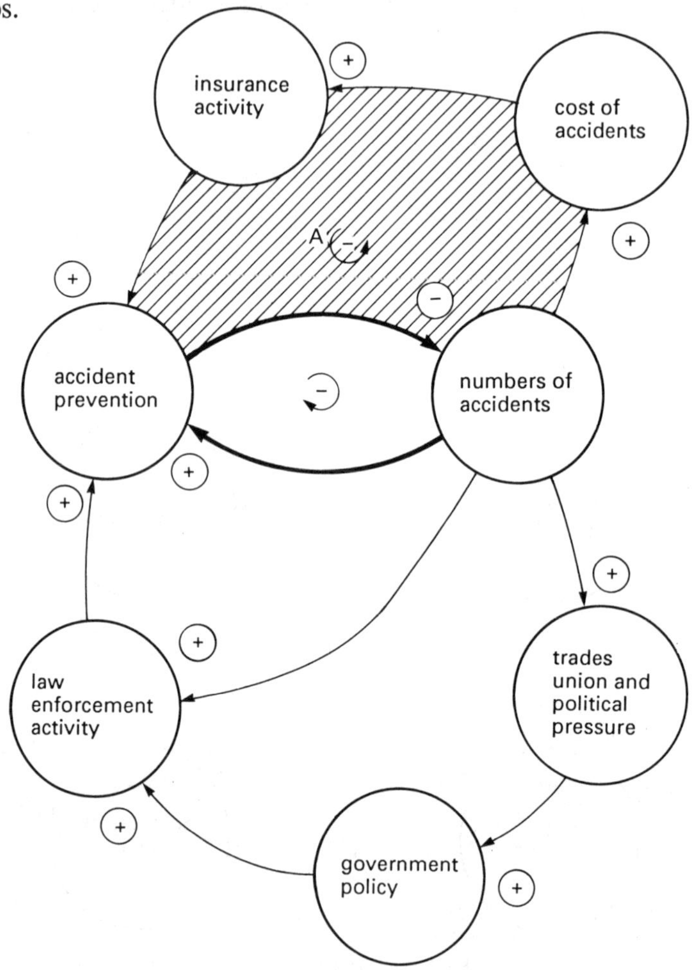

Fig. 2.10 Causal loop diagram of processes involved in accident prevention (digraph convention).

Exercises

7. In Fig. 2.10, give your interpretation of loop A (shaded area).

8. It is generally recognised that inflation leads to an increase in wage demands; successful wage demands in turn increase the costs of production and fuel inflation. This is, of course, a very naive explanation. Wage demands are stimulated by inflation and in particular by many other factors such as house prices. The demand for housing, availability of mortgages, movement of skilled people, and availability of jobs are other factors. Draw a causal loop diagram to help explain the economic dependency of people, housing and jobs.

Advanced panel

Generally, control loops that are negative are classified as adaptive. In other words, the system tends to respond to changes in the loop variables by counteracting those changes for stability. Typically, biological systems are self-maintaining or homeostatic, i.e. they tend to resist and adapt to changes in their environment. Control loops that are positive are classified as non-adaptive. They tend towards either rapid increase or rapid decrease. Such cascades result in system instability, for example uncontrolled money supply in the economy leading to hyper-inflation.

Causal loop diagrams are at the heart of simulatory models used to predict the effects of changes in particular variables (components). They are used widely by economists, engineers, biologists and other specialists and are often computerised. However, even highly sophisticated models are still only rather pale shadows of complex behaviour they are intended to represent. Note how we used qualified language in our descriptions above: 'tendency to limit' rather than 'it will limit'. So long as the limitations of simulatory models are borne in mind, the analyst can get a lot of value out of such models. However, fall into the trap of regarding the model as near-perfect and you will run the risk of blaming reality for failing to fit the model when it is the model that is not accounting for all aspects of reality!

Flow charts

A flow chart is a summary means for expressing any well-defined set of operations for solving a particular problem. Such a set of operations is

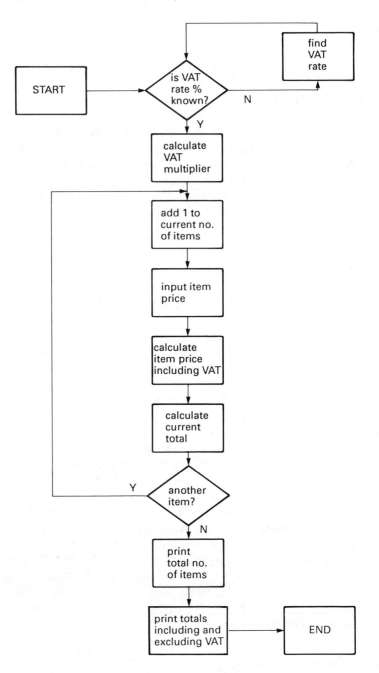

Fig. 2.11 A simple flow chart for adding up a VAT bill (error traps omitted).

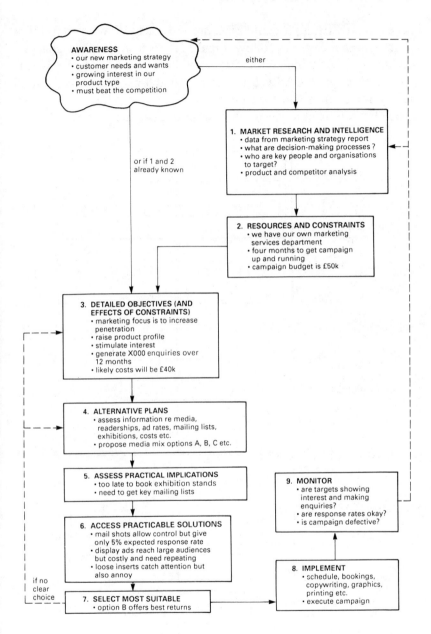

AWARENESS
• our new marketing strategy
• customer needs and wants
• growing interest in our
  product type
• must beat the competition

either

or if 1 and 2
already known

1. MARKET RESEARCH AND INTELLIGENCE
• data from marketing strategy report
• what are decision-making processes ?
• who are key people and organisations
  to target?
• product and competitor analysis

2. RESOURCES AND CONSTRAINTS
• we have our own marketing
  services department
• four months to get campaign
  up and running
• campaign budget is £50k

3. DETAILED OBJECTIVES (AND
   EFFECTS OF CONSTRAINTS)
• marketing focus is to increase
  penetration
• raise product profile
• stimulate interest
• generate X000 enquiries over
  12 months
• likely costs will be £40k

4. ALTERNATIVE PLANS
• assess information re media,
  readerships, ad rates, mailing lists,
  exhibitions, costs etc.
• propose media mix options A, B, C etc.

5. ASSESS PRACTICAL IMPLICATIONS
• too late to book exhibition stands
• need to get key mailing lists

9. MONITOR
• are targets showing
  interest and making
  enquiries?
• are response rates okay?
• is campaign defective?

6. ACCESS PRACTICABLE SOLUTIONS
• mail shots allow control but give
  only 5% expected response rate
• display ads reach large audiences
  but costly and need repeating
• loose inserts catch attention but
  also annoy

8. IMPLEMENT
• schedule, bookings,
  copywriting, graphics,
  printing etc.
• execute campaign

if no
clear
choice

7. SELECT MOST SUITABLE
• option B offers best returns

Fig. 2.12 Product promotional campaign as a decision sequence diagram.

called an algorithm; it could be expressed solely in words but a flow chart makes it easier to follow. Figure 2.11 is a simple flow chart for adding up a VAT bill. With a few modifications, it could be used to write a program in BASIC for a programmable calculator or point-of-sale terminal, although nowadays such flow charts have been superseded by other techniques in the computing world.

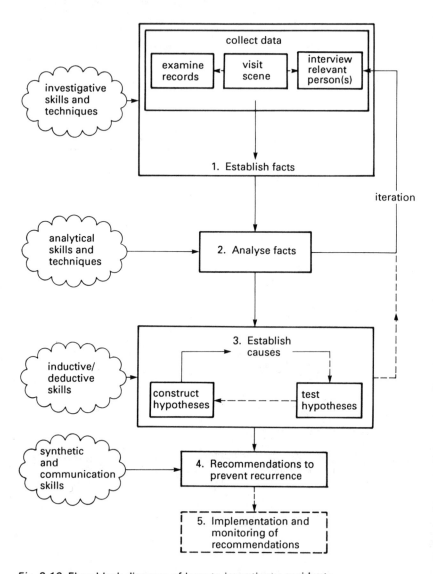

Fig. 2.13 Flow block diagram of how to investigate accidents.

Activity

Flow chart algorithms appear in many guises: tax return forms, instructions in telephone kiosks, passport application forms, etc., where written procedures have been at least partly turned into a flow chart. Find another example of a written procedure and convert into a flow chart.

Decision sequence diagrams

A decision sequence diagram is a special kind of flow chart which relates to rational choice decision making. The contents of each box or blob relate to choices or decisions and the connections between them represent either a logical sequence and/or actions. Figure 2.12 shows a decision sequence diagram for a product promotional campaign.

Flow block diagrams

Flow block diagrams depict practical procedures for doing something. For example, Fig. 2.13 summarises the sequence of things an accident investigator would have to do. Each square or block contains actions. The cloud blobs indicate skill resources that would have to be input.

Data flow diagrams

As the name suggests, data flow diagrams depict the flow of information in an information system. Figure 2.14 (taken from *Structured Systems Analysis and Design Methodology* by Geoff Cutts) shows data flow relating to order processing.

2.3 Preparatory description of the system

We have already impressed upon you the need to avoid jumping to conclusions such as 'this is an X problem requiring an X solution'. However, latching onto what is already known is much more comforting than delving into murky, uncharted waters. As any researcher, particularly in the social sciences, will tell you, dealing with uncertainty and the unknown can be stressful. As self-maintaining systems, humans tend to avoid the unknown unless they have the

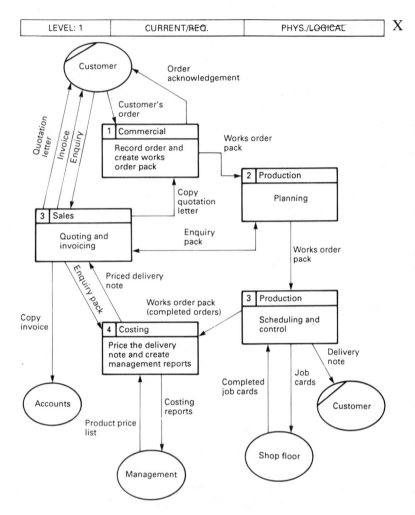

Fig. 2.14 Example of a data flow diagram. (*Source: Structured Systems and Design Methodology* by Geoff Cutts, Paradigm Publishing 1987.)

capacity to withstand or adapt to new discoveries and experiences. Therefore, any tool that can cast some light on murky waters *before* the analyst's voyage of discovery begins is to be welcomed. The pre-analysis tool that we have in mind is system description. Apart from helping you keep your sanity, it enables you to identify some key systems for detailed analysis.

There are five stages of pre-analysis leading to detailed analysis as the sixth stage:

(1) awareness/consciousness (of problems/issues)
(2) your commitments
(3) testing (is analysis a good idea?)
(4) separation (of a few relevant systems)
(5) selection (of one or two key systems)
(6) detailed analysis (e.g. proceed to hard systems analysis).

To show how pre-analysis works, we are going to use the car park levy case. Imagine you have been called in as a consultant by the chairman of the council's Coordination and Policy Committee who is concerned that their proposed decision may have unforeseen adverse effects.

Awareness/consciousness

You have expanded your own awareness by listing topics and constructing a spray diagram (Fig. 2.5). However, these structural components tell nothing of issues or feelings in the community about car parking. The 'messy' situation can be more evocatively portrayed by a 'rich picture' as in Fig. 2.15. The rich picture technique is described in detail in Chapter 5.

Commitments

Your own commitment to finding out is clear: you are being paid as a consultant. The rich picture, however, allows you to identify relevant persons or role figures who might also be committed to finding out more because the proposals raise problems for them. For example the Chair of the Coordination and Policy Committee is definitely committed; a road safety campaigner would have a commitment, too, as would a traffic warden. You could identify others.

As a consultant, you would normally only be interested in your own and your client's commitments, i.e. in the car park case the Chair of the Coordination and Policy Committee. For study purposes, however, the analyst could construct a 'commitment statement' for other problem owners. Such a statement does not have to be literally what they have said but can be a mixture of fact, deduction and induction. The aim is to put yourself in their shoes and capture the *essence* of what someone in their role would be committed to. For example:

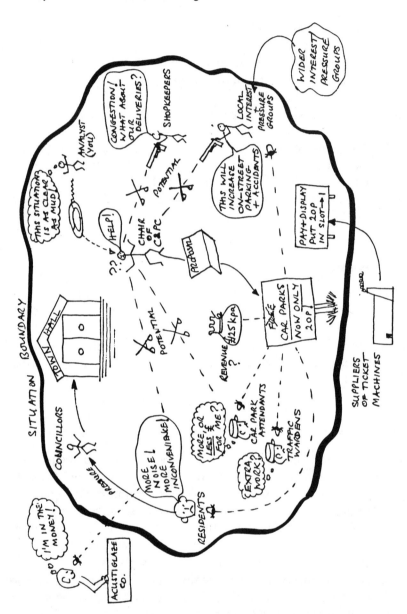

Fig. 2.15 A rich picture of the proposed car park levy situation as outlined in Fig. 2.5.

Chair of Coordination and Policy Committee: 'Our car parks are not paying their way. A 20p levy seems a modest charge and it will offset costs. But, I can foresee opposition from some quarters and I have nagging worries about unforeseen knock-on effects. The community expect us to get it right. I need to know far more about the likely implications.'

Road safety campaigner: 'Putting a levy on car parks is bound to increase on-street parking. The more cars and congestion on the streets, the greater the potential for accidents. We have a duty to improve road safety. I would like to know more about how the proposal came about and what evidence there is that road safety will not suffer.'

Traffic warden: 'On-street parking will probably increase. That means more illegal parking and more parking tickets to issue. Will we get extra resources to cope? I need to know if my job will suffer.'

Testing

Each commitment statement (and as a consultant you would only have your client's) is then tested to see whether it reveals sufficient content to warrant a systemic analysis. For example:

- will success be recognisable?
- is analysis the most effective way of getting results?
- would analysis be purposeful or simply indulging someone's idle curiosity?
- is the client's or the problem owner's goal important enough to warrant systems analysis?

Positive answers would be needed to each of these questions.

Separation

Separation involves firstly teasing out areas of concern from the 'successful' commitment statement(s). Thus, for example:

Different problem owners	Potentially fruitful areas
Chair of C&PC	A: traffic management B: road safety C: financial policy
Road safety campaigner	D: social behaviour of drivers E: council consultative processes

For each fruitful area, a trial definition and description of a system is set up. Thus, for area A a relevant system might be: 'The Council system for managing traffic on and off the highway.' The likely components of this system, both within the system and in its environment, are listed. This early separation reduces the uncertainty and nervousness felt by most analysts in setting the system boundary.

Within the system	*In the environment*
Chair, C&PC	Police
C&PC	drivers
councillors	public transport
highways department	residents
traffic wardens	ratepayers
car park attendant	shopkeepers
trade unions	local newspapers
car parks	Friends of the Earth
road markings and signs	RoSPA
parking meters	local campaigners
pay-and-display machines	
levies	
resources	

This process would be repeated for the other potentially fruitful areas. Each description would then be assessed to see whether as a potential system it is a key system, i.e. is it essential to making sense of the situation and to helping the problem owner's task? System A does appear to be a key system on both counts.

Selection

You may have separated out several key systems. The decision as to how many of them you carry forward for detailed analysis depends on your resources. Often only one or two are given the full treatment. You must use your own judgement as to which one(s) you select.

Activity

Separate another system from the list of fruitful areas (B to E) in the car parking situation; give it a title 'a system owned by ... for ...' and separate out its likely components; consider whether it would be a key system.

2.4 Summary

Diagramming is the major tool in systems work. Techniques are many and varied. However, such tools need to be used within a framework for describing and analysing systems. Systems description provides a preparatory framework before starting detailed analyses, such as those covered in later chapters.

2.5 Suggested answers to exercises

1. In addition to a title, every diagram should have the analyst's name and date on it. There should also be an indication of the state of iteration (see Explanation of Terms). It is not uncommon for diagrams to be developed through many iterations. How about your organisation chart from Activity 1?

2. An organisation chart of a human activity system usually describes only the *formal* structure and relationships, i.e. what it is supposed to be like. It gives no indication of informal aspects that usually are at least as important as the formal aspects.

3. Since the assessments sub-system overlaps customers in the environment, it suggests that customers are outside the system for the most part but are part of the system when one looks specifically at housing. For example, it determines who customers are, how they are defined, waiting list criteria, etc.

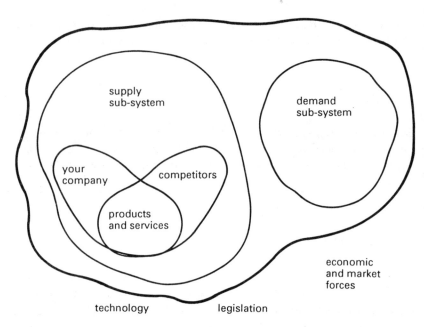

Fig. 2.16 A system map of the management training market (first attempt).

4. Usually, housing departments do not deal directly with external suppliers on new build but arrange for the architects and planning department to get the work done for them. The architects and planning department may do so through its own construction workforce (the DLO – direct labour organisation) and/or through external consultants and contractors. In contrast, usually housing departments organise their own maintenance and improvement work with or without the assistance of external suppliers. Thus, in Fig. 2.4 the maintenance and improvement sub-system reaches across to the external suppliers.

5. HM Government through legislation, grants, benefit rules, etc. takes precedence over and limits what a housing department can do. Figure 2.4 shows this by HM Government overlapping the housing department system boundary.

6. Our suggestions are shown in Figs. 2.16 and 2.17.

7. As the number of accidents increases, so does the cost of accidents. As the cost of accidents increases, insurers (who foot a large part of the bill) increase their activity. With an increase in insurance pressure on policy holders (e.g. premium increases), accident prevention increases. As accident prevention increases, the numbers of accidents decrease. Loop A is therefore a control loop.

8. Our causal loop diagram for the economic dependency between people, housing and jobs is shown in Fig. 2.18.

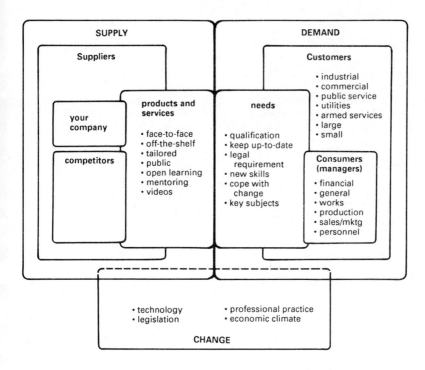

Fig. 2.17 A system map of the management training market (after several iterations).

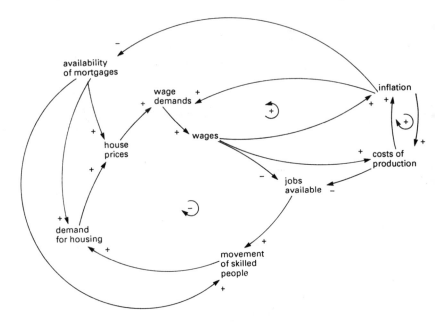

Fig. 2.18 Causal loop diagram of the economic dependency of people, housing and jobs. (*Note:* diagram incomplete and assumptions are debatable.)

Chapter 3
Introduction to Hard Systems

Overview

Hard systems are characterised by well-defined structures and processes and readily quantifiable features, all of which aid prediction and control.

Engineered systems are those that incorporate hardware or equipment. Natural systems include biological and other physical systems that are not man-made. Abstract systems are symbolic systems such as languages and signing systems.

Quantifiable measures of system efficiency include those that measure levels and those that measure rates.

World-views must be taken into account but are not a dominant feature of hard systems analysis.

3.1 Introduction

In Chapter 1, we showed briefly how to classify systems as either hard or soft. Hard systems may be regarded as those in which human behaviour is perceived to play a minor role, even though many people may in fact be involved in the system. The adjective 'hard' does not imply systems that are difficult to understand, although they often are. Rather, 'hard' refers to quantifiable, predictable and undisputed attributes such as when someone says 'give me the hard facts'.

This chapter opens up the hard systems concept and gives some graphic examples of hard systems.

3.2 Engineered systems

An engineered system is one that usually incorporates hardware or

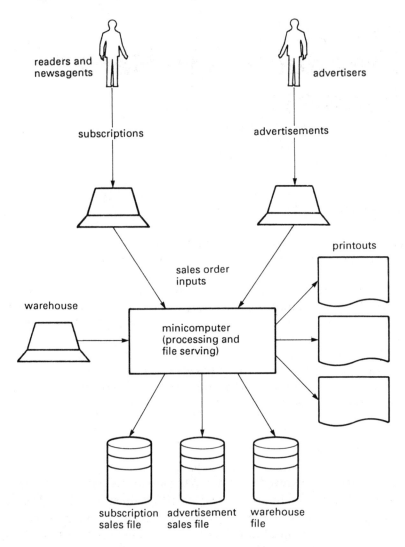

Fig. 3.1 Diagram of computer system in a popular magazine publishers.

equipment of some kind and has been designed to achieve certain desired goals. The central heating system in Chapter 1 (see Fig. 1.2) is an example. People may also form part of such a system to a greater or lesser extent, but usually it is taken for granted that they are only functionally important. For example, the driver is functionally important to the effective operation of a car (a small self-driven passenger transport system?) but the driver's world-view is considered to be largely irrelevant to the car's design and efficient working. Similarly, computerised data processing and information systems are usually designed in functional terms. Those who make the computer systems work are often assumed to be little more than functional appendages who, if they feature at all, can be reduced to single matchstick figures on systems diagrams. Such assumptions may be unwarranted. Although world-views are not central to hard systems, they should *always* be considered throughout any analysis.

Figure 3.1 shows a data flow diagram of a typical computer system in, say, a newspaper or popular magazine publishers. Figure 3.2 shows a diagram of a metal power press system. In the latter diagram, there is both flow within the power pressing process (e.g. blanks, power) and flow of information about the process (e.g. quality control).

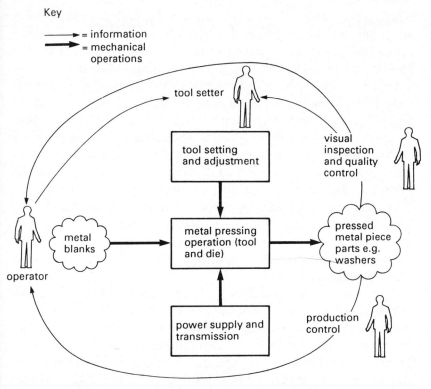

Fig. 3.2 Diagram of a metal power press system.

Exercises

1. Where would you place the system boundary in Fig. 3.1?

2. What kind of printouts (outputs) would you expect in Fig. 3.1?

3. In Fig. 3.2, what kind of feedback would production control be giving the operator?

4. Assuming the system boundary envelops the whole of Fig. 3.2, what components would you expect to find in the system environment?

3.3 Natural systems

Natural systems are those that are not man-made, for example biological systems, the weather, the oceans, and volcanoes. Some apparently natural systems are really engineered systems because they are designed, operated and controlled by man, e.g. a trout farm. The life cycle of the trout, however, is part of a natural system.

Other natural systems include micro-organisms. Controlling the spread of infectious diseases in man, animals and plants is a major concern of all governments. Such concern extends beyond humanitarian issues. If the population is debilitated by disease, for example, it will be less able to work efficiently to produce goods and so a whole economy may be weakened. A weak economy, in turn, may be unable to fund a health service capable of effectively combating disease. Thus, failure to control the spread of an infectious disease may lead to a 'snowball effect' (an epidemic) which in turn may lead to instability in non-biological systems. The fact that these kinds of problem lend themselves to quantitative modelling explains their inclusion in the hard systems area.

A characteristic of epidemics is exponential growth, i.e. the number of new cases reported for each successive period continues to increase. Thus, the total number of cases and the number of new cases per unit of time (i.e. rate of growth) are important measures of an epidemic's progress. However, such measures do not offer a clue as to the causes, transmission and spread of the disease. In order to predict and then control an epidemic's progress, you would need to know all the disease vectors (i.e. the dynamic variables) and their rates of change that characterise the epidemic.

Exercises

5. Intravenous drug abusers who have become infected with the AIDS viruses are said to be a key vector in the spread of AIDS. Identify other dynamic variables thought to contribute to the epidemic and draw a plausible causal loop diagram to simulate the epidemic.

Of course, many microorganisms are relatively harmless. Man has used natural systems such as yeast colonies for thousands of years in the production of bread and wine, for example. In this century, microorganisms have been used on a large scale to produce antibiotics. More recently, third generation biotechnology is being used to synthesise high purity organic chemicals on an industrial scale. In all these examples, man has been able to harness natural systems as part of his own engineered systems. Such control requires the ability to predict what the natural system will do under defined conditions. Prediction requires detailed knowledge of the dynamic variables that affect growth and product yields. For example, yeast growth depends on the temperature and acidity of the medium. It also depends on the number of cells in the colony and their age. From causal loop models and numerical data, it is possible to construct a numerical model which simulates the system's behaviour. As the numerical values of dynamic variables are changed, the model predicts the numerical values of the product yields, for example. If those predictions match what are measured in reality, then the model is reliable and has great practical value in production management. An example of numerical modelling is given in Chapter 4.

3.4 Abstract systems

Abstract systems are those created by the human mind. They have no physical reality although humans may create physical representations of them and they may be incorporated within engineered systems. Abstract systems may also be called symbolic systems; they comprise symbols organised by rules. For example, a natural language such as

English comprises vocabulary and punctuation (symbols) organised by the rules of English grammar. Mathematics with its symbols and rules is another abstract system. Commonly accepted symbols and rules confer predictability in interpretation.

Exercises

6. In what way are abstract systems important to the conduct of human affairs?

Computer languages and programs (lists of instructions) written in them are also abstract systems. They represent a means of communication between humans and computers. The example in Fig. 3.3 is part of a stock records program written in COBOL (Common Business Oriented Language) according to the rules of that computer language. In plain English, when the program is run the Stock File is opened and 'Which part number?' is displayed on the computer screen. The user keys in the part number. If the part number is invalid, 'No such record' is displayed. Otherwise, the user is instructed to 'Type quantity

```
PROCEDURE DIVISION.
      OPEN I-O STOCK-FILE.
      PERFORM UPDATE.
      CLOSE STOCK-FILE.
      STOP RUN.
UPDATE.
      MOVE SPACE TO PART-NO.
      DISPLAY "Which part number?".
      ACCEPT PART-NO.
      READ STOCK-FILE INVALID KEY MOVE "U" TO FLAG.
      IF FLAG = "U"
            DISPLAY "No such record"
      ELSE
            DISPLAY "Type quantity withdrawn from stock"
            ACCEPT AMOUNT
            SUBTRACT AMOUNT FROM QTY
            REWRITE STOCK-REC INVALID KEY DISPLAY "Error".
```

Fig. 3.3 Part of a stock records program written in COBOL. (*Source:* Dr Mike Oatey, South Bank Polytechnic; also appears in *COBOL from BASIC – a Short Self-Instructional Course*, by Mike Oatey and Carl Payne, Pitman, 1986.)

withdrawn from stock'. The computer then subtracts that amount from the item's stock quantity before closing the Stock File.

3.5 Quantification

Quantifiable measures are the information source used to predict and control the behaviour of hard systems. They are thus vital in determining whether the system is operating as its owners and operators intended, i.e. is success being achieved? People often use such measures of performance in a taken-for-granted way or may even fail to monitor them at all. Common examples are car fuel consumption (miles per gallon or litres per 100 kilometres) and domestic electricity consumption (kilowatt hours). The latter measure is a level of consumption whereas the former is a rate. Both levels and rates may be used to ascertain whether a system is functioning efficiently, i.e. whether the system's desired outputs are matching expectations for a given level or rate of inputs. If an electricity bill is higher than expected, it may be simply a case of greater use. However, greater use of electrical heating may have been stimulated by faults or lack of control in system components or sub-systems, e.g. heat loss from a poorly insulated building, or badly adjusted boilers and central heating units.

Often, a number of different measures are available for particular kinds of hard system. With computers, for example, a key structural measure is the amount of memory or storage capacity available to a user, either in main memory (RAM) or in backing store (floppy or hard disk). Computer storage is measured in bytes, one byte being equivalent to one character input from the keyboard. A byte is a small unit and so storage capacity is usually measured in thousands (kilobytes – Kb) or millions (megabytes – Mb). However, system efficiency is usually measured by a series of speed 'benchmarks' – how fast the computer system processes data. The performance of large computers is often measured in mips (millions of instructions per second). Although modern computers process data rapidly, a computer with a poor benchmark can result in perceptible and sometimes lengthy delays during processing. Whatever benchmark is chosen, it must be appropriate for the system.

Computer programs, too, can be designed with the result that they are less efficient in the way they run than they should be. A poorly designed program can take hours to sort or search through a file of records that a well-written program would tackle in seconds or minutes.

Exercises

7. For the following systems, suggest suitable measures of performance: ventilation system, a business quoted on the Stock Exchange, an hotel, a company PR department, a chemical reactor system.

3.6 Summary

Hard systems have characteristics such as clear structures and well defined processes that are readily measurable. Such quantifiable attributes enable a system's behaviour to be predicted, monitored and controlled. The world-views of people who own or operate hard systems must be taken into account but are usually not considered to be of *central* importance. The use of hard systems ideas implies a particular view of the nature of problems as explored further in Chapters 4, 9 and 10.

3.7 Suggested answers to exercises

1. The boundary would go around all the components but between 'readers/newsagents' and 'subscriptions' and between 'advertisers' and 'advertisements'. Readers and advertisers are in the environment. It is not unusual to see data flow and other system diagrams presented without a system boundary.

2. Obvious printouts would include invoices to customers (newsagents, individual subscribers and advertisers), sales ledger, stock lists, despatch lists, address labels. This diagram does not show various other likely functions such as purchase order inputs, and invoices received and payable. It also does not show what software (programs of instructions) would be used in processing. Note 'software' and 'soft systems' are entirely different things and should not be confused.

3. Production control would be relating the quantity of washers actually produced per shift to the quantity required. If the output is too low, then the operator may be required to increase the rate of inputting blanks. This quantification of performance is characteristic of hard systems. Quality control may also include quantified measures of quality, e.g. numbers of misshapen washers per 1000. Production control, quality control and tool setting are all part of this system's monitoring and control sub-system.

4. Typical components in the environment would be: production management, stock control, finance, accounts, payroll, personnel, sales, marketing, safety department, trades unions, customers, Power Press Regulations etc. Of these, production management represents the wider system of the metal power press system.

5. Other dynamic variables are: use of contaminated needles, infected homosexuals, infected heterosexuals, infected bisexuals, infected haemophiliacs, infectious sexual activity, infected blood supplies. Figure 3.4. shows their relationship as a causal loop diagram at first iteration.

    At present, the greatest prevalence of AIDS cases including HIV infection is among male homosexuals, bisexuals and intravenous drug abusers, and these categories together represent a tiny fraction of the population. However, as Fig. 3.4 shows, the worry is that the majority heterosexual population will become infected. Since sexual activity is a normal and frequent human function, an epidemic would then be difficult to control; at present there is no cure for AIDS. Figure 3.4 is of course inadequate. For example, no distinction is made between AIDS and HIV infection; no account is taken of educational and other efforts that might modify human behaviour and introduce control loops.

6. Many abstract systems are vital to human communication. Failure to use the correct symbols or to follow the rules may result in communication failure with possibly dire consequences. National and international conventions and agreements seek to minimise such problems, for example road traffic signs and signals and hazard warning pictograms on containers of dangerous substances.

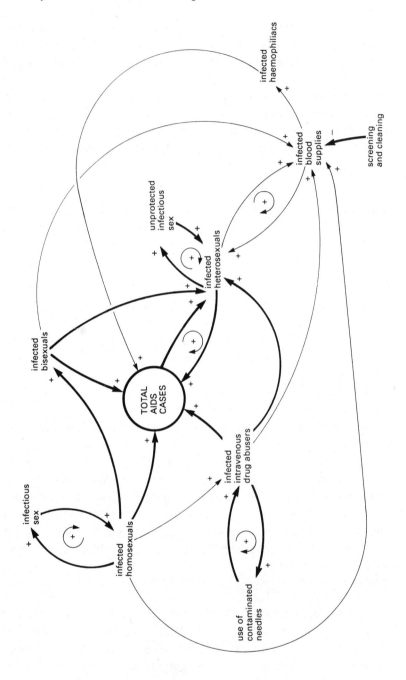

Fig. 3.4 Causal loop diagram of the AIDS epidemic.

7. Suitable measures are:

   Ventilation system: air changes per hour, linear flow rate (metres per second), volume flow rate (cubic metres per second).

   Business quoted on Stock Exchange: earnings per share, net profit before tax, stock turnover, profit as percentage of sales, return on capital employed (%), ratio of current assets to current liabilities.

   Hotel: room occupancy rate, staff turnover, net profit before tax, stock turnover, profit as percentage of sales, return on capital employed (%), ratio of current assets to current liabilities.

   Company PR department: enquiries per 1000 mail shot letters, feature articles placed per month, product write-ups per month, enquiries per display advertisement inserted, enquiries per 1000 pounds of advertising expenditure, sales conversions per enquiry (%).

   Chemical reactor system: per cent theoretical yield, tonnes product per hour, market value of product/cost of reactants, power consumption per tonne of product.

Chapter 4
Using Hard Systems Ideas

Overview

A hard systems approach to problem solving or opportunity fulfilment assumes that:

- problems/opportunities are identifiable, describable and soluble in a reductionist way
- there is an optimal solution or one that is clearly superior to others
- measures of performance are quantifiable
- systems models depend largely on mathematical relationships
- there is a large measure of agreement among the client set as to the nature of the problem/opportunity
- the client set largely agrees about the overall goal
- the client set's perception of the problem/opportunity is informed by a shared world-view.

The client set comprises all those with whom the study seeks to gain credibility. It must include system owner and problem owner and any key figures who may have an interest in or who may be affected by the study's outcome. In some cases, the analyst, system owner, and problem owner may be a single person undertaking the study on his own behalf.

The nine stages to a hard systems approach are:

(1) Groundwork
(2) Awareness
(3) Goals and objectives
(4) Strategies and options
(5) Measures
(6) Models
(7) Evaluation
(8) Making a choice
(9) Implementation.

4.1 Introduction

A wide range of examples of hard systems was given in Chapter 3. This chapter examines briefly the application of hard systems ideas to formal problem-solving and design activities.

4.2 The nature of problems

A large part of human endeavour is directed at solving problems and at constructing practical 'things' to meet unfulfilled needs. Most formal problem-solving activity involves a tacit set of assumptions on the part of problem solvers, namely:

- the existence of the problem may be taken for granted
- the structure of the problem can be simplified or reduced so as to make its definition, description and solution manageable
- reduction of the problem does not reduce effectiveness of the solution
- an optimal or superior solution exists
- selection of the optimum solution is a rational process of comparison (measures of performance against criteria).

The formal problem-solving procedure may be expressed as a decision sequence diagram as in Fig. 4.1. Although such a formal *procedure* is obviously systematic, it is not really a systems approach. A hard systems view of problem solving, while accepting most of the assumptions listed above, makes a more detailed and probing examination of the *system* experiencing the problem. Whereas formal problem solving starts to focus early on finding a solution, a hard systems approach does not.

An important question to ask is 'who says that X is a problem?' A problem does not actually exist outside an individual's mind; you cannot pick up a problem as you would, say, a cup of coffee. Thus, it cannot be taken for granted that a particular problem exists for what one person regards as a problem may be thought of differently by other people. Some may regard it not as a problem but as an opportunity to fulfil a need or achieve a goal. Others may have no particular opinion about it.

A hard systems approach to problem solving requires the analyst to purposefully check that there is a large measure of agreement among

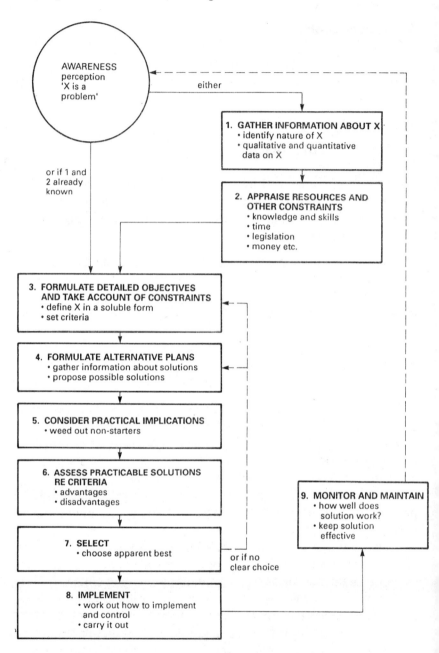

Fig. 4.1 Formal problem-solving procedure as a decision sequence diagram.

the 'client set' (see later section on analyst–client relationship) as to what the problem or opportunity is. For hard systems analysis to be effective, there will also have to be a large measure of agreement about the overall goal. This does not mean that everyone has to agree literally about everything; any group of people is likely to hold a range of opinions about matters of fact and to have differing interests. For example, the interests of the marketing department may be different from those of personnel. However, a fundamental assumption of hard systems analysis is that differences in *values* do not form part of the client-set's perception of the problem or opportunity. In other words, the analyst has to check early that there is a shared Weltanschauung about the present situation, the nature of the problem or opportunity, and the future situation (the goal to be reached).

Figure 4.2 is a simple model of problem/opportunity perception. Assuming that there is general agreement about the present and desired situations, the analyst's task becomes one of devising ways of getting from S_{now} to S_{fut} (i.e. strategies to reach objectives) and deciding which is likely to be the most effective option.

Activity

For a situation you are familiar with where change is being discussed or planned, examine the world-view of those involved for agreement about the nature of the problem/opportunity and the goals to be reached.

4.3 Analyst-client relationship

The analyst does not have to be an external consultant called in as an expert to provide a solution. As stated in earlier chapters, this is a DIY guide and there is no reason why the hard systems analyst should not be a manager or adviser within the organisation. The analyst's client set is one or a number of people for whom the systems study is undertaken but not necessarily those to whom the analyst reports. Remember that the system owner, the problem owner and the client set may be all different or all rolled into one. Indeed, the analyst can undertake the study on his own behalf. The client set may be defined as all those with whom the study seeks to gain *credibility*, for if the study and its results lack credibility among key decision makers and implementers then proposals for change may be rejected.

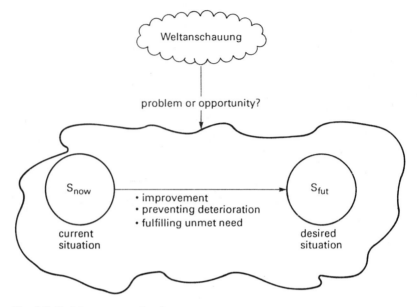

Fig. 4.2 Problem perception.

As part of the initial groundwork on problem/opportunity perception and gaining awareness and understanding of the system, the analyst has to decide who the client set is, what their interests are, and what their shared world-view is. This is important because the analyst's definition of the overall goal and the objectives to be reached must reflect those of the client set. The analyst needs to accept the client set's world-view, otherwise there is no point in continuing the study. However, the analyst may have good reason to modify the objectives and the client set may well be expecting guidance. The early stages of a hard systems study will probably involve some negotiation between analyst and client set towards an agreed set of objectives.

4.4 Hard systems analysis

Looking back at Fig. 4.1, you can see that formal problem solving is a relatively straightforward procedure. It can be used where problem solving is felt to require more thought than simply relying on past practice, experience or intuition, e.g. because the consequences of a poor solution may be serious. However, it is fairly prescriptive and, because of its assumptions, leaps quickly into 'devising solutions'. Where the problem is complex or involves many interwoven variables a more powerful method may be needed. Hard systems analysis provides such a tool.

A variety of hard systems methods is described in Section 4.6. However, hard systems methods share a common approach to problem solving that has evolved through cross-fertilisation of ideas and practice stemming from such experts as Ackoff, Churchman, de Neufville, Stafford, DeMarco, Yourdon, and the Open University Systems Group. The approach comprises the following stages:

(1) Doing the groundwork (identifying the client set and its world-view, and establishing communication and mutual confidence – see above).
(2) Gaining awareness and understanding of the problem (the current position; systems description – see Chapter 2).
(3) Establishing overall goal and set of objectives (the position to be reached; constraints to be contended with). Note 'goal' represents the overall target whereas 'objective' is a *measurable* contribution to the goal.
(4) Finding ways to reach objectives (creative, divergent thinking followed by structured focussing on a range of practical possibilities).
(5) Devising assessment measures (quantitative and qualitative measures of performance).
(6) Modelling (techniques to test possible options against measures of performance).
(7) Evaluation (assessing the likely outcomes of each option under a range of possible conditions; testing credibility with client set).
(8) Making a choice (selecting the route that best meets the objectives given the constraints and prevalent world-view).
(9) Implementation (putting the solution into effect; may require further systems design work).

The nine stages of this approach are not serial; some may be carried out simultaneously. As with all systems work, there will be iterative loops as later stages highlight defective understanding, woolly objectives, inadequate measures of performance, etc. Figure 4.3 summarises the approach.

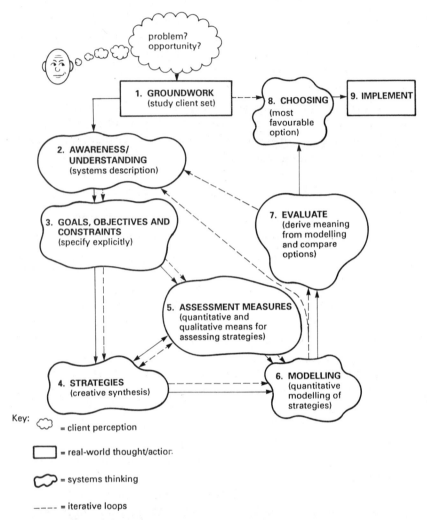

Fig. 4.3 Hard systems analysis for decision support. (Adapted from: John Hughes and Joyce Tait, *The Hard Systems Approach*, Open University T301 Course Material, 1984.)

4.5 Mini case studies

We have chosen two rather different examples to demonstrate some of the principles and stages of the hard systems approach. Neither case presents a complete run through of the method and we have been selective in the aspects chosen. Further coverage of the method and more extensive case studies are given in Chapters 9 and 10.

Tech-Abs Publications

Tech-Abs Publications Ltd specialises in producing weekly reports and bulletins comprising sets of abstracts of articles from scientific and technical journals. The company subscribes to some 1200 journals covering medicine, pharmacy, veterinary practice, plastics, engineering, etc. Articles are selected for abstracting by technical abstractors. Abstracts are typed on abstracting forms according to guidelines; a maximum length of 1200 words plus title and classification details. Abstracts are then edited and details checked before passing to the production department.

Tech-Abs has a high proportion of science graduates in all departments and most of the senior managers have worked their way up through departments from junior positions. A large multinational corporation now has a controlling interest in Tech-Abs and is exerting its influence, e.g. seeking greater professionalism in business strategy, increased efficiency, etc.

The market for abstract bulletins has been growing rapidly. Tech-Abs has, over 25 years, built up a commanding position in the market with its weekly bulletin service. Typical subscribers are large organisations throughout the world who wish to monitor not only technical developments in their field but also the activities of competitors and to spot market opportunities. Particular subscribers are increasingly demanding more specialised abstract services, e.g. only articles relating to particular classes of psychotropic drugs, or only articles relating to the design of direct broadcasting satellites.

Until 1983, batches of abstracts were sent daily to typists working at home who would re-type the abstracts on master sheets. A smaller number of internal typists also typed up master sheets and corrected those from the external typists. The master sheets were then available as camera-ready copy for the printers. Prior to being sent to the printers, the master sheets were passed through an OCR (Optical Character Recognition) scanner which created corresponding computer files for subscribers requiring them as an extra service.

The experience with the OCR scanner was bad; files were frequently corrupted and a lot of the computing department's time was diverted to trying to sort out one mess after another. The sales department were having to field an increasing number of complaints from dissatisfied customers. In any event, customers were beginning to demand a complete on-line computer searching facility whereby they could dial-up abstract files and search for abstracts according to whatever their current need was. For example, a customer might want to find all abstracts originating between January 1983 and January 1985 on substituted benzodiazepine psychotropics published in West German journals. Such precise searches could not be done on the OCR files.

There was general agreement among marketing, sales, production and computing departments that a better system was needed to improve production efficiency, to make it easier to produce printed bulletins tailored to individual subscriber needs, and to provide a fully searchable on-line database (computerised abstract files) that was reliable.

Exercises

1. Imagine you are in Tech-Abs management services department and have been asked as a systems analyst to help solve the problem. Who is the client set?

2. Does the current position represent a problem or an opportunity? Explain your answer.

3. Summarise the client set's world-view in so far as you can judge from the information provided. Are there other world-views that need to be considered?

Figure 4.4 was produced by one of the client set to summarise the production position as she saw it. This helped the analyst at the systems description stage. The overall goal was to introduce flexibility by adapting the current system of data handling which was dominated by manual methods (typing, hard copy corrections, hard copy masters etc.) so that the abstracts could be captured electronically at an early stage in the publishing cycle. Once stored in a reliable computerised form, the abstracts could be searched, sorted, categorised and used quickly and in a wide variety of ways.

A number of objectives were identified such as:

- speed up entry of abstracts onto computer
- introduce computer-searchable indexes on each abstract
- speed up error corrections

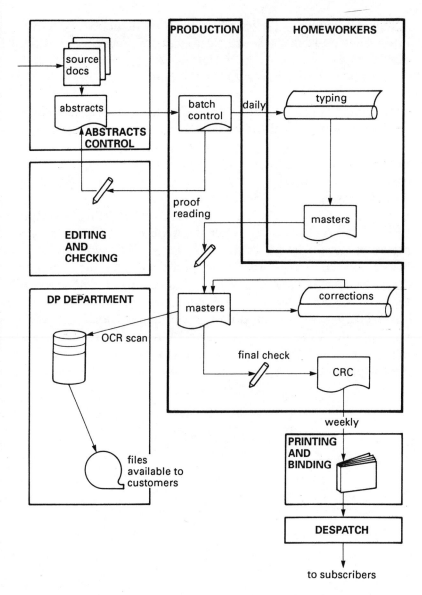

Fig. 4.4 Flow block diagram of Tech-Abs production run superimposed on system map of departments (1983).

- maintain current weekly publishing schedule and 4-weekly production cycle
- individualise bulletins for customers.

Figure 4.5 shows the objectives in the form of an objectives tree or hierarchy. Constraints included the need to introduce changes without disrupting a tight publishing schedule and production cycle. Any new system that required substantial staff training would be likely to be disruptive.

A number of possible routes were considered. One was to have the abstractors key their abstracts directly into a microcomputer, thus removing the need for typists and two keying operations. This was rejected, however, because the abstractors were not 'computer literate', would require training, and had indicated a reluctance to change their way of working.

The route that was eventually chosen involved the replacement of the typists' typewriters by microcomputers and a special word processing package. Keyboard entry of abstracts was faster than before and the typists-turned-word processing operators could produce 1000 abstracts plus searchable indexes on disk in the same time it took to type 1000 abstracts. Corrections could be done quickly on-screen before any hard copy printouts were made. These corrected files on floppy disks were then dumped into a main file on the company's main computer. Once a week, indexed files were passed to a phototypesetting computer which printed out camera-ready copy

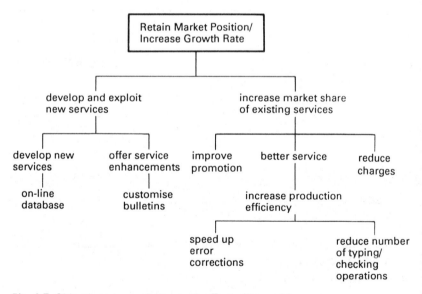

Fig. 4.5 Objectives tree relating to the Tech-Abs problem.

according to subscriber requirements. The main indexed database file was now also available for on-line searching by subscribers. Figure 4.6 shows the production system some two years after the implementation of the revised system (as summarised by the production manager).

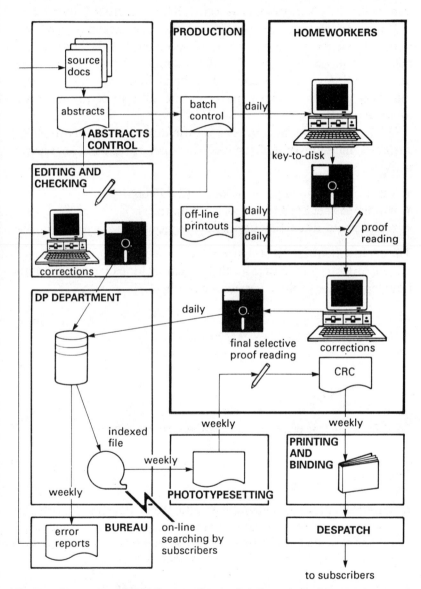

Fig. 4.6 Flow block diagram of Tech-Abs production run superimposed on system map of departments (1985).

Superstores

Superstores PLC is a chain of supermarkets. Like all retail chains, Superstores experiences a phenomenon known as 'shrinkage', i.e. loss of stock through theft by customers and staff. Superstores estimate an annual loss of 1% of stock turnover which currently is £300 million.

You as Superstores' security adviser have been asked by the board to investigate the problem. On your first iteration of the hard systems method, you deliberately keep things simple. As you gain greater understanding of the system and its behaviour, you can afford to introduce greater complexity into your measures of performance and modelling.

While considering how to reduce shrinkage, you identify three key variables (measures of performance) that operate regardless of security activity:

- value of stock in any one month S (£ millions)
- shrinkage rate SHR (£ millions per month)
- replacement rate REP (£ millions per month).

The replacement rate is a figure added to normal turnover through sales in order to compensate for theft. A causal loop diagram shows the interdependence of these variables (see Fig. 4.7).

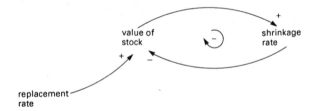

Fig. 4.7 Causal loop diagram of Superstores' shrinkage problem.

Exercises

4. From the causal loop diagram and the three measures, construct a numerical model that relates the stock value in any one month to the effects of shrinkage and replacement. Assume that the starting stock value is S_0 and the monthly time interval is T.

5. If the starting value of stock for the first month S_0 is £20m, use the model from Q4 to calculate the stock value after the first six months.

Assume that the shrinkage rate is £0.25m per month and the replacement rate is £0.20m per month, i.e. failing to keep pace with losses.

The cumulative *discrepancy* after 6 months is £20m – £19.7m or £300 000 but of course this is deceptive because the actual *loss* is 6 × £0.25m or £1.5 million. The model highlights the practical problems of stock monitoring, for theft is instantaneous whereas identification of losses and their replacement always lags. It may be difficult to completely match losses by replacements. Although a large part of losses are insurable, they still represent an unacceptable drain on assets and eventually insurers will demand higher premiums based on Superstores' claims history.

Clearly, something needs to be done to reduce the shrinkage rate. You have been considering a range of security options but before examining each one in detail you want to get a feel for what effect a security provision of so-many £ per month would have on the situation.

Exercises

6. Expand the causal loop diagram (Fig. 4.7) to include security input.

A very simple numerical model to show the likely kind of relationship between the shrinkage rate (SHR) and the security variable (SEC) is SHR = k/SEC where k is a constant. Unsurprisingly, this new model suggests that an investment in security will reduce the shrinkage rate. For example, if the current level of security investment (SEC) is £10 000 per month and the shrinkage rate is £250 000 per month, the model suggests that raising SEC to £20 000 per month would reduce shrinkage to £125 000 per month.

Of course, the simple model SHR = k/SEC is crude and therefore unreliable; the relationship in the real world is not a simple inverse proportion. Although increasing the security investment would probably have a desirable effect, it is highly unlikely that doubling the security investment would literally halve the shrinkage rate. It is known that shrinkage is almost impossible to totally eradicate and there will be a level of SEC beyond which SHR does not fall appreciably, i.e. the law of diminishing returns. Other factors would therefore have to be taken into account in a refined model, e.g. quality of security package, quality of its implementation, level of security maintenance, changes in social conditions.

4.6 Different hard systems methods

We have used the term 'hard systems approach' as if it were a singular method. Historically, two branches of hard systems have grown in parallel: decision support (DS) and systems engineering (SE). Both employ the general framework of hard systems analysis as outlined above and in Fig. 4.3 but their emphases are different. The DS approach emphasises problem solving at strategic and tactical levels in organisations. It provides a rationalistic tool for mapping a way forward from the current position. Typically, problems and opportunities concern marketing strategies, business development, product development, etc.

Whereas a DS study may conclude that a new computer system with a Local Area Network (LAN) is required, systems engineering emphasises design of such technical systems. SE translates decisions into working systems; for example, practical options to reach objectives would have to consider several LAN configurations that are available, their specifications, advantages and disadvantages, and their costs. In the Superstores case, a DS study might have concluded with a decision to invest a sum of money in a specified security package; an SE study might then have been warranted to convert the decision in principle into an effective *operational* security system.

In practice, hard systems studies often incorporate aspects of both DS and SE at the same time. A lot depends on the problem. The Tech-Abs case provides an example of both DS and SE considerations.

There are a growing number of proprietary hard systems methods. Usually, these have been developed for application to particular kinds of problem. The computing and information technology fields in particular have spawned a large number of methods which are mostly variations on a theme, for example:

JSD	Jackson Structured Design
SSADM	Structured Systems and Design Methodology
SASD	Structured Analysis and Systems Design
YSM	Yourdon Structured Method

Further reference to such methods is made in Chapters 10 and 17. In the production and mechanical engineering fields, SE methods include:

MRP-II   Manufacturing Resource Planning
PLOT   Production Logistics Organisation Technique

Once you have grasped the ideas presented in this book, you should have little difficulty in learning the characteristics of these variants.

4.7 Summary

It could be argued that both the Tech-Abs and the Superstores problems could have been solved without recourse to hard systems analysis. In the Tech-Abs case, for example, it is easy to see that direct data entry onto computer is very appropriate. The two case studies were greatly condensed and simplified as a means of introducing a number of aspects of good practice that should typify hard systems analysis (e.g. consideration of world-view, objectives hierarchies, numerical modelling). In particular, hard systems analysis seeks to avoid the pitfalls of solving the effects of *symptoms* rather than the *causes* of the symptoms.

Examples of knee-jerk solutions that merely treat symptoms are countless. Just one will suffice as a reminder. Sales were dropping in a car dealership. The solution tried was to install bigger, better and brighter showrooms. This failed because the owner did not appreciate that the main cause of poor sales was the dreadful reputation they had for after-sales service, i.e. an emergent property of that dealership as a system.

Compared with formal problem solving, hard systems analysis provides a greatly enhanced approach to solving complex problems that are readily structured and that can be reduced to a quantifiable model. The approach is examined in more detail in Chapters 9 and 10. However, the hard systems approach has limitations as discussed in Chapter 5.

4.8 Suggested answers to exercises

1. The client set must include departments which have already expressed a clear interest in the problem, namely marketing, sales,

production and computing. You may decide that only the heads of those departments are key decision makers but you may well decide that the study will also have to gain credibility with some of their subordinates. The client set may also have to include other key figures not yet mentioned, for example those who will have the power to allocate or withhold financial or other resources, the managing director who has to keep within policy guidelines laid down by the parent company, and so on. You must decide who the client set is.

2. The client set sees the current situation as a problem that needs to be resolved (poor OCR performance, wasted time of computing staff, customers' needs not being met). However, routes to objectives might be devised that not only solve the current problem but also create opportunities, for example the creation of new products and services.

3. Scientists tend to be fairly sober and rationalistic in their approach to things, and the client set is dominated by science graduates. They are not likely to favour flamboyant or 'risky' solutions. Tech-Abs is a company that rewards dedication and loyalty, as evidenced by senior managers having worked their way up; the client set cares about the company, the products and the customer. Therefore, the objectives and routes to them will have to be feasible within that culture that characterises Tech-Abs in general and the client set in particular. They are likely to value an improvement in the existing system rather than to replace it with a radically new system.

   Of course, in describing the system it is not just the world-views and commitments of the client set that you will have to make explicit. What is *your* world-view and commitment? Do they accord with the client set's?

4. Given that:
   - stock value at time T months is S_T
   - stock value at the start is S_0
   - replacement rate is REP (£m per month)
   - shrinkage rate is SHR (£m per month)

   the equation that models their relationship is:

$$S_T = S_0 + T(REP - SHR)$$

$$
\begin{aligned}
5.\ S_6 &= S_0 + 6(0.2 - 0.25) \\
&= 20 + 6(-0.05) \\
&= 20 - 0.3 \\
&= \pounds19.7 \text{ million}
\end{aligned}
$$

6. The expanded causal loop diagram is shown in Fig. 4.8.

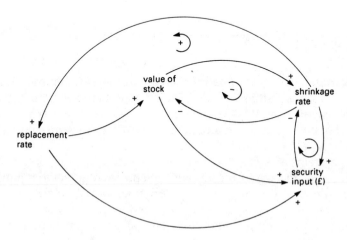

Fig. 4.8 Expanded causal loop diagram of Superstores' shrinkage problem.

Chapter 5
Introduction to Soft Systems

Overview

Soft systems concern human activity and especially hard-to-quantify aspects such as attitudes and relationships between one group of people and another. When such human activity systems exhibit crisis, conflict or unease, it is inappropriate to use a hard systems approach to try to solve 'the problem'.

Soft systems analysis starts with a rich picture or situation summary which serves to capture the various, and probably somewhat jumbled, mental images you have of the unstructured problem situation. Structures and processes may be depicted by any convenient means. Although a *situation* boundary may be marked, a rich picture is not a *system* diagram.

5.1 Introduction

You will have learned by now that 'hard' in our sense of the term does not mean 'difficult'. Similarly, 'soft' does not imply systems that are easy to deal with – usually quite the reverse. With hard systems, it is often assumed that human beings form only part of the system in a simplified and predictable way – as if they were simple inanimate components of a machine or other engineered system. The features of hard systems are readily capable of measurement and quantification. Their behaviour is relatively easy to understand and to predict.

5.2 The nature of soft systems

Of course, in many ways human beings are anything but predictable in the way they behave. Each individual is unique, not only physically but also in the way they think, act and feel about things. Each person's view of the world (or Weltanschauung as described in Chapter 1) is made up of a complex set of attitudes, beliefs, values, opinions and perceptions. Each of us reveals only a glimpse of his or her own world-view in our relationships with others – and this may lead to misunderstandings and conflicts. Each of us is constantly guessing at what other people *really* mean, what they *really* intend, and *really* believe. When the General Manager announces his intention to reorganise a department, some people likely to be affected may read into the announcement all sorts of ulterior motives that may or may not be correct. In human relationships, it is perhaps easy to see how disputes and strikes start, how marriages break down and how wars break out! Part of a manager's job is to anticipate misunderstandings and to aim for better communications, but even with the best of efforts sometimes 'wicked', 'messy' situations may result.

Soft systems are those concerned with human activity of some kind and the Checkland method described in this book is specifically intended for use where a human activity system exhibits crisis, conflict, uncertainty or unease in relationships among the human 'actors'.

5.3 Lucrative Publications – a mini case study

The following description of Lucrative Publications (based on a real case) illustrates a typical soft 'problem situation'. Whatever your views on the uses and abuses of English, in soft systems it is conventional to refer to a problem *situation* rather than the idea of 'the problem to be solved'. The reasons for this subtle and important difference will become clear as you proceed through this chapter.

The unstructured problem situation

Background facts about Lucrative

Lucrative Publications Ltd is a small company of about 100 staff based in Central London and specialises in publishing yearbooks and reference handbooks for professional and trade associations. The company has existed for 15 years and has always been located at the same four-storey converted Victorian premises. Originally, the company was owned by three brothers who were ex-advertising space salesmen.

Twelve months ago, the brothers sold out to an entrepreneur who wanted to move into publishing.

The company ethos prior to the acquisition was that of a conventional small commercial publisher, namely maximisation of advertising revenue coupled with minimisation of overheads and production and sales costs. Total staffing had been held at about 40 to 45 for some five years, the nominal breakdown being 30 telephone sales people, five sales administrators, five accounts and payroll staff, five typists, and the three brothers as publishers and editorial directors.

The company's policy was to pay outside subject specialists to act as consultant editors rather than to employ full-time editors. Production editing, layout and paste-up were all handled by a freelance in Peterborough. The printers were in Gloucester. The editorial content typically comprised a mix of informative articles and reference sections since experience had shown that readers wanted this. The viability criterion for publishing any book was the projected ratio of advertising pages to editorial pages. However, provided that a 'reasonable' balance was reached publication would continue. With some books a cover charge would also be made.

About two-and-a-half years ago, the company's gross advertising revenue began to level off. The number of readers willing to purchase books also began to fall. The recession was blamed. The brothers decided to sell the company to someone who would inject capital and who had ideas for expansion.

The new owner as Managing Director began to 'sweep clean' quickly. The major changes were:

(1) The appointment of two full-time editors to handle list-building, commissioning, text editing and production (so replacing the consultant editors and the freelance production editor).
(2) The introduction of computers into the accounts section and the replacement of typewriters by VDU word processors. No additional typists or operators were appointed and no special training was thought necessary for the changeover; the new editors were also provided with VDU word processors.
(3) The expansion of the sales department with the appointment of a 'dynamic' Sales Manager and a doubling of the telephone sales staff; the new owner decided that advertising revenue was still top priority and the Sales Manager was given authority to overrule editorial decisions if he felt it expedient, e.g. he could substitute editorial advertisements (revenue) for commissioned articles (costs).
(4) Physical reorganisation of the staff and offices so as to accommo-date the new staff; the premises became even more cramped than

beforehand; there was a rumour that the company might move to larger premises in Northampton some 80 miles away.

Signs and symptoms of organisational distress

Ever since the introduction of the VDUs there were complaints from the typists-cum-VDU operators and the new editors that VDUs were dangerous and that their health was being damaged. A long list of alleged illnesses had been logged yet investigations by expert ergonomists and safety specialists failed to substantiate any of the allegations. Still the allegations persisted and staff relations fell to a low ebb. The word processing operators said that they were being forced by the editors and sales staff to work at sweat shop rates on equipment that was dangerous. The editors said that they were being forced to work as skivvies for the sales department and, to cap it all, they were having to work with hazardous equipment. Recently, several of the word processing operators and editors joined different trades unions in order to pursue their fight against using VDUs. In addition to the company's internal problems, purchasers of the books have for the first time complained about the quality of editorial content. The Managing Director is in despair about what is going wrong and has called you in as a systems consultant.

Exercises

1. Justify describing the Lucrative setting as a 'soft' problem situation rather than a 'hard' problem.

5.4 Beginning your analysis – the 'rich picture'

We now come to one of the most delightful aspects of soft systems work – the rich picture or situation summary. Rich pictures can also feature in systems failures work and so you will meet them again in following chapters.

To show how to construct a rich picture, let us take one of the centres of apparent conflict in the Lucrative scenario, namely the editors vs. the advertising staff. The editors' grievance has already been outlined but words can be much more evocatively portrayed in picture form. Figure 5.1 depicts how we see the conflict. The cartoon (rich picture) captures the *essence* of the conflict between the editors and advertising staff and it readily conveys the story to others. The old cliché 'a picture is worth a thousand words' really does have some merit.

Fig. 5.1 A rich picture of the conflict between editorial and advertising staff at Lucrative Publications.

Exercises

2. Take another centre of conflict in the Lucrative story – the word processing operators' grievance – and construct your own rich picture. Remember, artistic talent is of little importance here whereas some imagination is necessary.

Suppose you wanted to create a rich picture summary of the whole Lucrative scenario. Obviously, both the sketches developed so far would feature in it. To create a rich picture of the whole setting, get a large sheet of drawing paper and 'plonk' down the pictures conjured up in your mind's eye by the description you have. The result, probably after several improvements, should be a mixture of 'structural components' or things that are relatively stable in the setting such as editors, Sales Manager, Managing Director, accounts section, and 'processes' or activities, transient relationships, and connections of some kind as in Fig. 5.2.

Your own rich picture of Lucrative might only bear only a passing resemblance to ours. If so, it does not matter. Remember, it is *your* understanding of the situation that you are trying to capture. Note that you can use whatever symbols you find convenient. The table in Fig. 5.3 lists some of those that are often used.

Notice in Fig. 5.2 that we have drawn a boundary around the situation. Purists may argue that rich pictures must not include such a boundary. We are less dogmatic. As you know from previous chapters, the setting of boundaries is important in all systems work. Please note, however, that a rich picture is *not* a system diagram – it is a visual summary of the human activity situation that you are concerned with at the *start* of your enquiry. The reasoning of soft systems thinking assumes that the sheer complexity of human behaviour will defy superficial attempts at rationalising. Thus, by setting down a *system* boundary on a rich picture you would be rationalising (i.e. trying to make sense of it according to previous knowledge) which at such an early stage would be self-defeating. You would be super-imposing and locking your thinking into a prejudicial framework that suggests 'answers'. Some politicians may be able to get away with 'instant judgements' and 'instant solutions', if only because they may soothe public anxieties. You, however, are seeking effective and long-term improvements. So, if you choose to draw in a boundary make sure that it is not a *system* boundary.

The style of your rich pictures reflects your personality and your world-view in general. Those having vivid mental imagery tend to draw them on the rich side like those in this book. Bear in mind, however, that you may wish to show your rich pictures to other people and especially actors in the problem situation. You can avoid possible offence by having two versions – one for your own reference and a cleaned-up version for showing to other people.

Advanced panel

Where does the information come from that indicates a soft problem and from which you can construct a rich picture? Within an organisation there may be many sources, for example reports, memoranda, minutes, interview notes, meetings and chance encounters that in varying

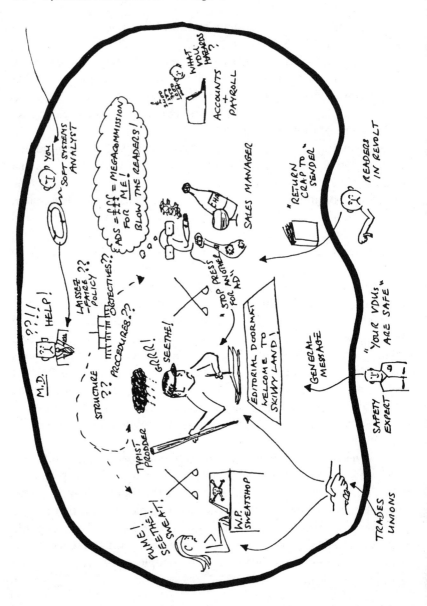

Fig. 5.2 A rich picture of the 'mess' at Lucrative Publications (at first iteration).

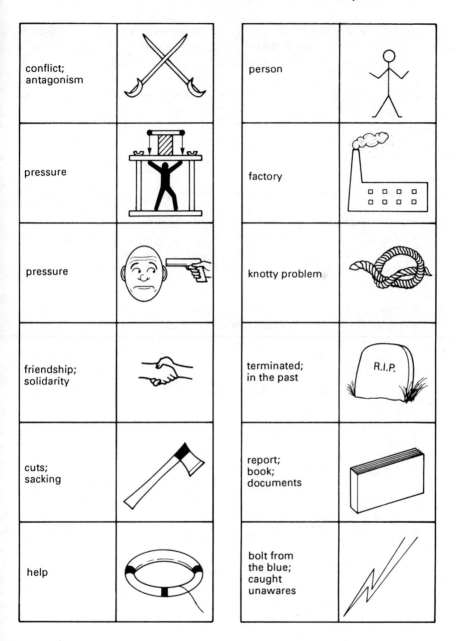

conflict; antagonism		person	
pressure		factory	
pressure		knotty problem	
friendship; solidarity		terminated; in the past	R.I.P.
cuts; sacking		report; book; documents	
help		bolt from the blue; caught unawares	

Fig. 5.3 A selection of visual symbols for use with rich pictures.

degrees betray disharmony or dysfunction. Interviews with a selection of the actors are really a necessary requirement but many people are ill-equipped to conduct an interview to best effect. Just because most of us have been talking to people all our lives does not mean we are all good at interviewing! Knowing how to approach people, how to phrase non-directive questions and how to respond to answers are most important. We cannot provide a detailed description of interviewing technique in this book. Some training videos on the market concerning recruitment and selection offer some useful guidance on interviewing technique.

Some hints on interviewing technique

As an interviewer, you need to establish rapport at the start. Rapport is a state of mutual confidence and harmony. Talking briefly and casually about neutral topics such as the weather and sport is one way of developing rapport before getting down to the subject of the interview. In summary:

- make sure you have got the right person
- interview on a 1:1 basis wherever possible
- avoid interruptions (telephones, visitors etc.)
- establish rapport
- clearly state your purpose however diplomatically it is wrapped up
- ask open-ended questions (why was that? could you tell me more about ...?)
- avoid superimposing your own biases on questions and responses
- get interviewee to do most of the talking
- make notes
- avoid interrupting to clarify (leave till later)
- check through your notes at the end and then clarify points
- have you forgotten anything?
- thank interviewee for assistance
- close interview but mention possibility of follow up if there are any points to clarify.

Newspaper reports and journal articles may also be a useful source of initial indications, especially if you are hired as an external consultant to the organisation in question.

Exercises

3. Read the newspaper report in Fig. 5.4 on the problem situation concerning air traffic control. Identify clues to the soft nature of the situation. Construct a rich picture of this situation.

Call for air traffic inquiry

**By Michael Smith,
Industrial Editor**

Air traffic controllers are demanding an urgent independent inquiry into the system after a report on declining staff morale, frequent equipment breakdowns and near-misses among aircraft.

The report, published yesterday, states that morale is poor among 79 per cent of controllers at the UK's busiest air traffic centre.

Mr Chris Stock, president of the Guild of Air Traffic Controllers (Gatco), the professional body representing almost 1,500 controllers, said : " The time is now ripe for an independent inquiry into the UK air traffic control operation."

The controllers plan to press

the Transport Secretary, Mr Paul Channon, for an inquiry into civil and military air traffic and because of the increasing concern about the safety of the system.

Concern about the air traffic control system will have increased after the publication of a survey among 214 controllers at the London traffic centre at West Drayton, Middlesex. The survey, which has been analysed independently by the applied psychology unit at Cranfield Institute of Technology, catalogues declining morale, complaints about equipment, and worries about the quality of management by the Civil Aviation Authority.

The report says there is " some cause for concern "

where so many staff are experiencing some degree of disaffection with their work. " The results clearly indicated that there is a concern among air traffic controllers about the standard of equipment with which they work."

Only two of the 214 controllers who filled in questionnaires for Gatco's survey had not experienced a failure of the West Drayton computer system in the past six months.

The Institution of Professional Civil Servants, which represents many of Britain's air traffic controllers, responded to the report by declaring : " Controllers have no confidence in their management or equipment."

Mr Bill Brett, assistant gen-

eral secretry of IPCS, said : " The controllers lack confidence in the CAA because they are not consulted on air space, equipment, or even their own jobs." The institution is pressing the CAA to provide more air space for commercial flights to ease congestion and introduce simulators at West Drayton for training.

The Civil Aviation Authority said serious attention was being given to staff morale, but Mr Keith Mack, controller of the National Air Traffic Services, added : " In view of the continuing management dialogue with the controllers, the guild's survey inevitably covers ground which is already very familiar to me and does not appear to reveal any new problems which are not already being tackled."

Fig. 5.4 Call for air traffic enquiry (*The Guardian*, 1 August 1987).

4. Repeat Exercise 3 for the following report concerning a dispute at a tobacco factory.

'While negotiations on their reinstatement founder, sacked workers at a Blogton cigarette factory have received backing from Lord Ted Grimbal. Six women were dismissed two weeks ago when they stopped work to support a colleague suspended for refusing to operate a new machine at the Blogton Tobacco Company. They have set up picket lines and have appealed to other workers not to cross until they have been reinstated.

The National Joint Council of the Tobacco Industry met yesterday and heard from both management and unions. It ruled that the six had been wrongly dismissed and that the two sides should negotiate their reinstatement. But management refused to budge and the meeting was adjourned for two weeks.

Lord Ted praised the 'principled' stand of the strikers and their supporters when he made a surprise visit to their picket line. 'I am shattered to learn of the breakdown in industrial relations,' he added. 'These workers are not militants or troublemakers. The grievance they sought to raise was legitimate. If they were to reinstate these six, the company would restore its previous reputation for fair play.'

Union branch secretary Jim Noakes claimed that management wanted to 'smash the union at the factory' and said that the six were paying an enormous penalty for a principle.' (*The Blogton Advertiser*, 1987).

5.5 The analyst's role

This is a convenient point to step back and consider your role as analyst. So far we seem to have taken it for granted that you are a neutral observer in the scheme of things – that you are like a scientist examining living organisms under a microscope. Such an assumption is open to challenge and indeed the soft systems method explicitly requires that you as analyst recognise your contribution to the setting you are investigating and the impossibility of 'neutrality' or true objectivity in your interpretations.

It is important to try to avoid passing judgement on the rights and wrongs of the situation of interest. We deliberately chose the previous two examples because they involve matters upon which many people hold strong views. For example, you may hold a general belief (really a 'value') that it is management's job to manage and that trades unions often create a lot of fuss about nothing. Alternatively, you may hold an opposing view that suggests that workers are inherently at a disadvantage and are vulnerable to exploitation by employers. Neither value is open to objective scrutiny and each is an artefact of the individual's world-view and its sources (see Chapter 1). There is no way of *proving* that management is destined to make all decisions of importance, any more than there is of *proving* that workers are entitled to a greater share in decision-making power.

Advanced panel

Most people do not question or probe their own set of values or the assumptions that underlie them. These values and assumptions are tantamount to systematic biases that influence the individual's thinking. There are two implications from this. First, to be competent in soft systems analysis (and in systems work generally) you must try to be brutally frank with yourself about your own values, assumptions and motives. Be explicit and include a statement on them in your reports so that others may judge what influence they may have had on your line of enquiry and its outcome. Second, you as analyst will both affect and be affected by the situation you are investigating. Your world-view through your behaviour will be interpreted by others and their behaviour will be modified as a result – and vice-versa. Where the problem is essentially soft in nature and you are, say, a manager in the situation who is attempting to analyse and remedy matters, you may find it very difficult to disentangle yourself from the 'mess'. As an actor already in the setting, you are as much a part of the problem situation as anyone else. You will be affected by corporate culture, customs and social interactions to an extent you may be unaware. You will have a stance towards and opinions about what is causing 'the problem'.

All of this immersion may cause you difficulty in 'seeing the wood from the trees' at two levels. At one level, there may be perceptual failure due to familiarity whereby work practices and social behaviour hallowed by the passage of time become almost ritualistic. Few people, probably including you, will be aware just how absurd it appears to an outsider. If confronted with a request for an explanation or justification, you would probably be reduced to comments like 'Well, it's always been done like that'.

Blind spots

The author was once asked by a large bureaucratic organisation to advise it on improving its health and safety policy and the organisation and arrangements for implementing it. After an initial assessment, it became clear that the problems they were experiencing were essentially soft in character. One particular incident exemplifies the problem of 'blind spots'. The author was shown an impressive-looking statement of safety policy, a legal document required under statute. Failure to draw up such a policy or to have an effective one may render an organisation and/or its officials liable to criminal proceedings. On perusing the three-year-old document, the author noticed a section headed 'Safety Committee' which set out in detail its composition, aims, frequency of meetings and so on. The minutes to committee meetings can be a good guide not only to historical development but also to issues and to general style and attitudes. On asking for copies of the minutes from the previous three months, the author was met with stunned then embarrassed silence. Eventually, the committee chairman confided that 'of course, the committee is still feeling its way and has not yet felt the time was right to have an inaugural meeting'. The actors (or at least some of them) had failed to notice the glaring disparity between what they had committed themselves to in a statutory document and what they were actually doing.

At a more insidious level, some blind spots may be deliberate and heavily value-laden, especially where they reflect vested interests that are being protected. Values (and vested interests) may be shared implicitly by one group of people and not by another. The gulf between the value sets and the actions they inspire may be such as to create issues – bones of contention that may fester and create the crises, conflicts, unease or uncertainties that characterise soft problems. Even in cases where issues appear to be obvious or have been stated, apparent explicitness may be misleading. It is common experience that people are naturally guarded and are unlikely to reveal the real reasons for their dissatisfaction or what they regard as the issues at stake. If expressions are made at all, they are likely to be muted or else 'dressed up' in coded language that is deemed politically safe and publicly acceptable. In the committee example above, 'feeling our way' and 'time not yet ripe' is coded language for 'we don't want to do it'!

Exercises

5. Read once again the newspaper report on the air traffic control problem situation. Identify some examples of coded language in the various quotations.

Imagine that you are the Data Processing manager of a company that is experiencing difficulties in introducing computers into its offices. You attempt to identify the causes of dissatisfaction by visiting users and discussing their problems. You discover that users are unhappy with 'unfriendly' software and manuals. It is relatively 'safe' for them to be frank about whether or not the computer and its software does its job and whether or not the manuals are understandable. However, they are most unlikely to reveal that their underlying dissatisfaction is with what they regard as the unwarranted power that your department exercises over them on such things as computer selection and access to computer facilities. As an inside actor in the setting, you are destined to remain ignorant of this issue unless you are a particularly sensitive and perceptive individual. Users may be more revealing to an outsider who may also be able to pick up clues that you would fail to spot.

The problems of seeing the wood for the trees and identifying worldviews, vested interests and issues that actors would prefer remain hidden present difficulties for any analyst. For the manager-analyst who is a working member of the situation he wishes to analyse, those difficulties are compounded by his familiarity with it and other actors' sense of self-preservation. These difficulties are of course relative but in contrast to the 'hard' and 'failures' approaches the use of an outside 'soft' analyst is often preferable.

5.6 Summary

In contrast to hard systems, soft systems explicitly concern behaviour in human activity systems and especially tangled webs of conflict, unease, misunderstandings and uncertainty that are difficult to unravel let alone 'solve' by conventional methods. A soft systems approach recognises as central the world-views of groups of actors and key figures in the setting. The soft systems analyst needs to be able to probe defensive behaviour of actors and to recognise and interpret coded language.

Rich pictures enable a summary of the mess to be captured. The use of this technique forms an early stage in the soft systems method described in the next chapter and in Chapters 11 and 12.

5.7 Suggested answers to exercises

1. The Lucrative setting is essentially soft in character for the following reasons:

   (a) Although there are apparently elements of both 'failures' (e.g. the possibility that the company had failed to protect its workers from hazards) and 'hard' problems (the possibility of demonstrable cause-effect relationships between VDUs and the alleged illnesses), these had been effectively dismissed by the earlier expert investigations.

   (b) The Managing Director is at a loss to know why there appears to be so much staff discontent and does not know what to do about it. This lack of clear objectives coupled with no general agreement within the company about what is wrong suggests that a hard systems approach would not be appropriate.

   (c) There are clear indications of overt discontent about the alleged VDU hazards and illness among the word processing operators and editors. However, there is also a clear suggestion that other more subtle and hidden values are operating. For example, the word processing operators (former typists) clearly feel that they are getting a raw deal as a result of the new Managing Director coming in and making them do much more work on the VDUs in cramped conditions and without proper training (e.g. expressions such as 'sweat shop'). The editors feel that their professional status and independence is being undermined (e.g. expressions such as 'skivvies for the advertising staff'). The fact that the accounts and payroll staff have *not* complained about VDUs may be significant.

(d) The current situation is not at all clear. There is a nagging suspicion that there are all manner of hidden undercurrents (in addition to those identified in (c) above) that may be relevant to the situation. For example:

- for whose benefit is the current policy on advertising and editorial content – the purchaser? the editors? the Sales Manager? or whom?
- what do the staff (the 'actors' in the situation) feel about the rumoured move to Northampton? are people worried about losing their jobs?
- there seems to be some personal antagonism between the Sales Manager and the editorial staff
- there seem to be numerous world-views that are not reconciled in the 'new' organisation.
- there is an underlying issue of control over job, tasks etc. which is focussed through the presenting problem of discontent about VDUs.

2. Figure 5.5 shows the word processing operators' grievance in rich picture form.

Fig. 5.5 A rich picture of the word processing operators' grievance at Lucrative Publications.

3. Clues to the soft nature of the situation: declining staff morale, increasing concern about safety, worries about quality of management, staff dissaffection, no confidence in management or equipment, lack of confidence in CAA, but management feels problems are already being tackled. The problem situation in rich picture form is shown in Fig. 5.6.

4. Clues to the soft nature of the situation: breakdown in industrial relations, escalatory spiral of action and reaction, entrenched views. The problem situation in rich picture form is shown in Fig. 5.7.

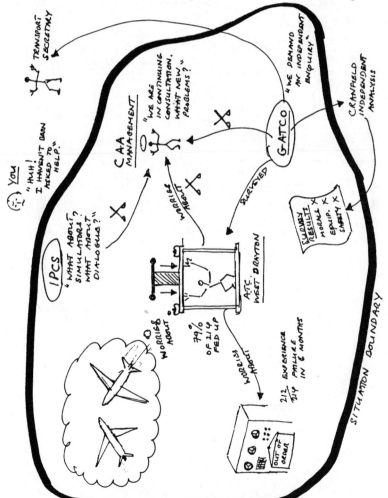

Fig. 5.6 A rich picture of the air traffic control problem situation (at first iteration).

5. Examples of coded language:
   'the time is now ripe' = 'it should have been done ages ago'
   'some cause for concern' = 'people are hopping mad'
   'survey covers ground already familiar to me and does not appear to reveal any new problems' = 'because I don't recognise any new problems everything must be alright'.

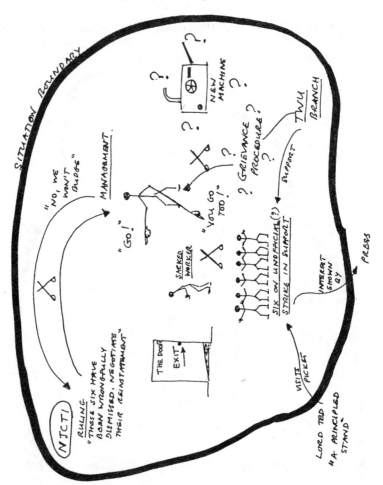

Fig. 5.7 A rich picture of the problem situation at the Tobacco Company (at first iteration).

Chapter 6
Using Soft Systems Ideas

Overview

The unstructured problem situation is summarised as a rich picture from which issues and primary tasks areas that appear relevant are teased out. This early analysis is of 'what is', i.e. the problem situation as presented in the real world.

Consideration of issues and primary task areas leads to selection of relevant systems and framing of root definitions. These must be hypothetical and not drawn from or offering real-world solutions. They relate to what 'might be' rather than what 'ought to be' or 'will be'.

The root definition of a relevant system is converted into a conceptual model showing the relationship between this hypothetical system's essential processes or 'verbs'. This model must not contain concrete examples from the real-world. Comparison of the model with the real-world problem situation enables an agenda of topics to be drawn up for discussion. Only if aspects of the conceptual model are agreed by the actors does consideration of practical requirements occur.

6.1 Introduction

Having established some of the characteristics of soft situations in Chapter 5, we are now going to have a quick run through the soft systems method. This method, pioneered by Professor Peter Checkland (see Useful Reading) and promulgated by the Open University Systems Group among others, provides a means of action for change in situations that are 'messy'. The method is capable of being used to tackle situations where the hard systems approach would be inappropriate and would be unlikely to produce expected results. The Checkland or soft systems method is represented diagrammatically in Fig. 6.1. The method differs markedly from hard systems approaches

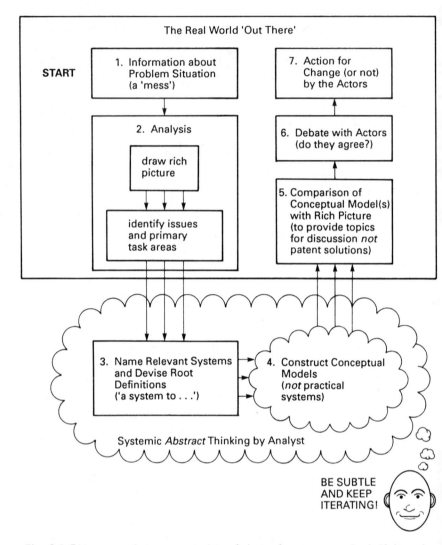

Fig. 6.1 Diagrammatic representation of the soft system method. (Adapted from P. Checkland, *Systems Thinking, Systems Practice*, John Wiley & Sons, 1981.)

in that it does not focus on finding a solution to a predefined, structured problem; the purpose is not to devise ways of reaching objectives. With soft systems, problem solving only comes at the *end* of a process of enquiry. To managers who are used to *starting* by focussing on problem definition, setting objectives and a creative search for solutions, the Checkland method may be somewhat disconcerting.

6.2 Issues and tasks

The Checkland method starts by collecting information about the unstructured problem situation or 'mess' and creating a rich picture as described in Chapter 5. Having drawn your rich picture, the next stage is to study it and try to identify parts of it that appear to be important or relevant to the overall problem situation. In searching for clues, there are some rules-of-thumb that are helpful as described below.

Look back at some of the cases and their rich pictures in Chapter 5. Examine a rich picture for symbols of conflict (e.g. crossed swords), someone under pressure (e.g. man in wine press) and uncertainty (e.g. question marks). These are usually good clues to stimulate your thinking. Also, do not stifle your own hunches that, for example, *apparent* harmony in part of the picture is really masking deep-rooted antagonism or mistrust that may be relevant.

So, where is all this leading? What *exactly* are you looking for? You are seeking to make explicit your perception of what is 'wrong' with the present situation and is causing so much trouble. In addressing yourself to what appears to be wrong, two kinds of 'wrongness' may be discernible – problems that are based in the primary task and problems that are issue-based. By primary task we mean the overall objective purpose(s) of the human activity system or part of it that you are examining. The primary task of a manufacturing concern is to produce marketable goods. The primary task of a furniture manufacturer, for example, is to convert wood into furniture. If he stopped doing this primary task, he would no longer be a furniture manufacturer.

In order to carry out a primary task, however, sub-systems may be required for:

- production
- provision of customer spares
- sales
- stock control.

You may have identified that, for example, the causes of dissatisfaction lie in the sales function not fulfilling one of its objectives such as meeting sales revenue targets. There is no obvious reason but your hunch is that the system for meeting that objective is in some way defective. You might then conclude that a 'system for maximising sales revenue' is a relevant system for further examination.

In a hospital plagued by organisational malaise, the staff might all express a commitment to the primary task of patient care (i.e. everything that is done to and for a patient). However, your observations suggest that some administrators actually behave *as if* the hospital is an 'administrator career benefit system'. In the Lucrative case, it appears that the Sales Manager regards the publishing house as a 'sales commission maximising system'. In other words, an apparent mismatch between the goals of different groups or individuals and their perceptions of the primary task is leading to disaffection among those with less 'clout' or assumed power.

Latching onto primary tasks is tempting because it deals with aspects of organisational life that are relatively concrete and with which you may feel more at home. Primary task systems are the stock-in-trade of traditional analysts. However, primary tasks generally duck the real issues and so may not always produce an effective outcome. For example, it may well be that a primary task system is defective *because* of unresolved issues.

An issue is something that evokes emotional responses such as frustration, anger, despair and resentment and which may lead to conflict or disaffection. For example, in the hospital mentioned above one issue might be the way in which management decisions are made that affect patient care. Another might be the role of administrators in ensuring patient care. Generally, you should try to tease out several issues and, perhaps, one or two primary task areas for consideration.

Whether particular problem situations are issue or primary task based is often debatable. For example, you may perceive the primary task of a hospital as being to cure sick people and that 'patient care' is a subsidiary welfare issue. If, however, you perceive patient care in an holistic way as indicated earlier, then patient care is the primary task.

Exercises

1. Identify some possible issues in the Lucrative story, the tobacco factory news item and the air traffic control news item as described in Chapter 5.

6.3 From analysis to systemic thinking

The next two stages of the Checkland method require a certain amount of mental agility. You will be thinking not in terms of 'what is' or indeed of 'what ought to be' or 'what will be', but rather of 'what *might* be'. Further, you will be thinking at a purely *logical* level of what your *hypothetical* system would *have to* contain if it were to be functional. Further still, the components of this system have to be framed *not* in terms of practical choices but in terms of *functions*.

The emphases have been deliberate as experience shows that many people find it difficult to think in the abstract. Thinking in concrete terms of familiar, tangible choices (e.g. container lorries) may be more comforting than abstract concepts (e.g. goods transportation). In the latter example, the choices of *how* goods are to be transported are back in the real world and should not be restricted as they would be in the foregone conclusion of container lorries. Such choices and decisions are left for a much later stage in soft systems analysis. The Checkland method deliberately avoids looking for solutions or strategies to objectives, in contrast to the hard systems approach.

So, at this stage try to clear your mind of 'solutions'. Your short list of issues and/or primary tasks are the seeds of systems that appear relevant to the problem situation in the real world. In the abstract world you are entering, your relevant systems will be fleshed out and tightly defined as 'root definitions' – the roots of *hypothetical* systems.

6.4 Relevant systems and root definitions

In the Lucrative case, one of your relevant systems might be entitled: 'an editorial and advertising reconciling system'. A first stab at the root definition of this system might be:

> 'a system to ensure that editorial decisions reflect the best interests of the publishing house'.

Now a closer examination of this root definition reveals some rather woolly and vague wording. What, for example, are 'the best interests'? 'Best' in whose terms? Who is operating the system? A second attempt is:

> 'a system to be operated by the Managing Director to ensure that editorial decisions reflect the objectives of the publishing house in

terms of financial viability and profitability and in terms of reputation in the marketplace among purchasers and advertisers.'

Notice how this first refinement or 'iteration' has made the root definition much less ambiguous although it still needs tightening up. The title too can now be modified to reflect more accurately its root definition: 'a system for making cost-effective editorial decisions'.

Exercises

2. One of the difficulties facing Lucrative is the growing dissatisfaction among book purchasers about quality or value-for-money. Write a root definition for a hypothetical 'system to satisfy the needs of customers'. Make one iterative refinement to your root definition.

Iterative development of the root definition knows no limits but obviously you would need to call a halt at some point. A good test for whether your root definition is adequate is the CATWOE test. This mnemonic stands for:

- *Customers*: (beneficiaries or victims of the system; not necessarily customers of the company)
- *Actors*: (those involved in operating the system)
- *Transformation*: (the essential process)
- *Weltanschauung(en)*: (world-view(s) of actors)
- *Owner(s)*: (power figure(s) who control the system; not necessarily owners of the company)
- *Environment*: (constraints on the system).

You have met some of these terms earlier in the book. The essence of the test is to see whether the CATWOE elements are explicit or implicit in the root definition and whether any omissions are critical. Notice that in line with abstract thinking the editorial decision making system dreamt up for Lucrative is subtle in concept. The transformation or essential process is *not* taking 'editorial copy' plus 'advertisements' and converting them into 'marketable books' as outputs. The essential process of our abstract system is editorial *decision making* – a process essential to Lucrative regardless of whether it were publishing books, boys' comics or electronic news-sheets and whether or not display advertisements were carried. 'Editorial copy', 'advertisements' and 'marketable books' are all concrete options from the real world. We shall be returning to the CATWOE test in more detail in Chapter 11.

6.5 Conceptual models

Having satisfied yourself with the adequacy of your root definition(s), the next stage is to construct a conceptual model of what each system *logically* would have to comprise in order for it to work. Remember that you are still working with hypothetical systems and not 'solutions' and so practical elements such as 'word processing facilities' or 'fleet of container trucks' must not appear in your model.

As an illustration of how you would develop a conceptual model from a root definition, let us take the root definition we have already drawn up from the Lucrative case: 'a system for making cost-effective editorial decisions'. The root definition itself does not include the essential components for such a system. Start by considering what the input(s) and output(s) of the system would be. Our suggestion is shown in Fig. 6.2. The essential activities or 'verbs' would be:

- *determine*: market needs and wants
- *formulate*: policy and objectives of publishing house
- *allocate*: resources
- *set up*: communication and control procedures
- *operate*: communication and control procedures
- *establish*: criteria for editorial decisions
- *decide*: on editorial questions
- *monitor*: performance.

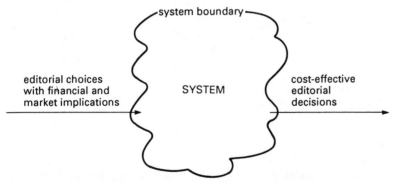

Fig. 6.2 The beginnings of a conceptual model for a system for making cost-effective editorial decisions.

The conceptual model now looks like Fig. 6.3. As you might expect, this is no more than a first attempt and is doubtless inadequate. In the proper use of the method, tests for adequacy would have to be applied

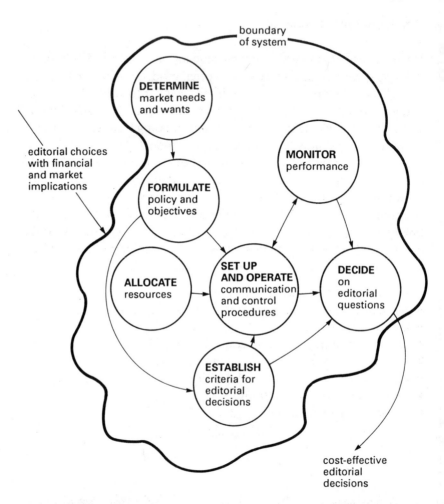

Fig. 6.3 A conceptual model of a system for making cost-effective editorial decisions (at first iteration).

and these are covered in Chapter 11. Each of the activity 'blobs' will need to be expanded to more fully describe the essential processes. Development of the model is iterative until you feel that you have gone far enough.

6.6 Comparing 'what might be' with 'what is'

Welcome back to the real world! You are now in a position to compare your conceptual model with your rich picture with the purpose of identifying discrepancies. You can do this in a number of ways ranging from intuitive inspection to more structured assessments. For example, you could examine in turn each activity in the conceptual model and check whether it does in fact happen in the real world situation. If it does not happen, why not? If it does, why? Ask yourself a lot of why? who? and what? questions. Alternatively, you could envisage the whole conceptual model as a system in operation and compare it with what actually happens.

The aim of the comparison is to arrive at a list of activities or topics that are either missing or perhaps unsatisfactory in the real-world situation. And, just in case you were beginning to relax after an uncomfortable journey through the abstract world and starting to think in terms of real-world 'solutions' – *don't!* Your list is really an agenda for discussion with the actors – you raise the list of activities that are apparently defective but it is up to *them* to decide how, if at all, they should be changed.

Do not pre-empt the final stage of the method by offering patent solutions. For example, in the Lucrative case various back-up activities to communication and control procedures may be missing such as apparent gaps in management structure, unclear division of responsibilities, lack of agreement on authority etc. These are items for discussion; the agenda should therefore not include statements like 'General Manager needed', 'needs an Editorial Manager', 'Sales Manager should have no editorial remit', and so on.

In common with other system methods, iteration is a permanent feature of the soft method. The comparison stage may well suggest that there are parts of the problem situation that remain opaque and so you may have to gather more information to sharpen up the rich picture. This in turn may cause you to reflect upon your relevant system(s), root definition(s) and conceptual model(s) and perhaps to modify them.

6.7 Discussing the agenda

You, as analyst, are not providing instant or even tailor-made solutions in the way that, say, a conventional management consultant might do. You will be suggesting some possible changes in the situation. Some, and maybe even all, of your suggestions may get the 'thumbs down'. Whatever the case, those that pass through for action will have to be valid in systemic terms and feasible within the culture of the organisation. For example, many large bureaucracies in the public sector have directorates that are run semi-autonomously by Chief Officers. A suggested change that would (or would appear to) upset the status quo and jealously guarded power would probably not be culturally feasible. In practice, you will probably have developed conceptual models from two or three root definitions and so will not have to rely on just one agenda for discussion.

Agreed changes typically will fall into one or more of the following categories:

> structural changes (organisational and/or physical)
> procedural changes (processes and activities)
> policy changes (overall objectives and strategies).

Sometimes it will be agreed that attitudinal changes are needed but it is notoriously difficult to try to engineer attitude changes. Education and training programmes may provide a practical method of raising *awareness* but attitude changes per se may have to be allowed to emerge. Attitudes both feed and are fed by cultures to which individual actors belong. The culture of a whole organisation or groups within it (i.e. unwritten and usually unadmitted rules of behaviour, ideologies, language, rituals etc. that characterise the set of people) is usually resistant to rapid change but may be coaxed into a gradual change. Such a change is more fundamental than others such as changes in equipment or work methods and some people may feel threatened – after all a culture change would undermine a lot of what they take for granted about their own identities. Considerable skill and commitment are usually needed to effect such a change.

6.8 Action for change

Specific actions, in terms of *how* to implement agreed changes, are matters for the actors. However, an agreed change may well point to a hard systems study as a desirable way of *planning* the change. For example, if an agreed change is 'an improved information processing facility' this might be achieved in a number of ways such as computerised data processing (various), microfiche, videodisk or even improved manual methods. A range of hard systems methods (e.g. SSADM) might be suitable for detailed planning and implementation of the most suitable change.

6.9 Summary

As a means of action for change, hard systems methods assume an early agreement among the actors on the nature of the problem and the goals to be reached; the exercise is largely one of devising ways to reach objectives, i.e. solutions. In contrast, the Checkland method starts with the assumption that there is not *'the'* problem to solve but rather a 'mess' – a system of problems that defies resolution simply by tackling each one separately (even if they could be identified and defined).

Action for change only occurs after a careful and systematic application of real-world analysis, systemic conceptualisation, comparision and debate. You, the analyst, are thus not a problem solver but someone who provides insights and facilitates others (the other actors) in understanding the differences between 'what is' and 'what might be'. In Chapter 12, you will be taken through case studies that illustrate soft systems analysis in detail.

6.10 Suggested answers to exercises

1. *Lucrative issues*: how editorial decisions are made; relationship between editorial and advertising staffs; the move to Northampton. *Tobacco factory issues*: industrial relations machinery; grievance procedures; trades union and management power. *Air traffic control issues*: reliability and safety of equipment; staff morale; communication between management and controllers.

2. A system owned by Lucrative which aims to satisfy the needs of purchasers of its range of books coupled with a return on investment to Lucrative commensurate with at least maintaining its share of the yearbook/handbook market.

Chapter 7
Introduction to Systems Failures

Overview

An apparent failure is defined as a shortfall between someone's expectation (i.e. 'success') and what they perceive to happen in reality. Dissatisfaction with something is implicit in perception of failure. In the systems context, failure is an emergent property of a system. Although systems failures that are investigated are usually past events, they may be current or even *potential* future failures.

The formal system model has to have the following sub-systems: control, performance monitoring, and operational sub-system(s). The system exists within a wider system that enables it to exist and both are affected by an environment. Characteristics of system failure shown up by comparison with the formal system are:

- deficiencies in organisational structure
- sub-systems are inadequately designed
- performance of sub-systems is deficient
- communications are deficient
- wider system fails to set objectives
- imbalance between operational resources and monitoring and control resources
- environmental disturbances underestimated.

7.1 Introduction

Previous chapters have introduced the concepts of hard and soft systems. Any system (hard or soft) is capable of failure. System failures can be dramatic and sometimes the consequences can be very serious. Study of system failures enables a great deal of value to be learned for the improvement of systems and the prevention of future failures.

7.2 What is a failure?

Most people would claim to recognise a failure when confronted with one. The car will not start; the company's latest product only meets a quarter of projected sales; the power failure that wipes all your data off the computer; the breakdown in pay negotiations between management and unions. Presented in this way, failures may seem obvious. However, just as systems and problems only really exist as ideas, so too do failures. You cannot touch a failure. You can only perceive something to be a failure, i.e. see it as a failure in your mind's eye. Just as 'hard' and 'soft' are used in systems terms differently than in everyday speech, so too is 'failure'.

Failure is a device of definition. What one person may regard as a failure may be considered as a success by someone else. At a personal level, if your car breaks down you are likely to regard this as a failure (e.g. mechanical failure, personal failure because you ignored the oil light flashing, servicing failure at the garage). However, suppose you were in the lead in a car rally – your rivals might regard your car breakdown as a success! Most people regard wars as failures, but arms dealers and black marketeers who profit from wartime economies see the breakout of peace as a failure. World-views and vested interests play a large part in perception of success or failure.

Perception of failure is fairly easy to detect. Expressions in conversation, in reports and in the media such as 'it's a failure', 'an absolute disaster', 'heads must roll', 'crisis must be resolved', 'what went wrong?' all betray a failure to achieve something that someone thought should have been achieved.

Exercises

1. In 1978, the Three Mile Island nuclear power station at Harrisburg in the United States suffered a LOCA (Loss of Coolant Accident). Sifting through evidence from official investigations, the following points emerge:

   - no major radioactive release occurred
   - there was disruption of normal life on a massive scale over a wide area
   - the plant safety systems proved largely effective
   - public confidence in nuclear power was severely damaged

- no one was injured
- the emergency was badly managed at every level.

In your opinion, was this a failure or a success?

Because a failure is a product of a person's perception it is always preferable to refer to an *apparent* failure. However, the more predictable and certain are the properties of a system, the more likely people are to agree about whether or not a failure has occurred. Failures of hard systems, especially engineered technical systems, fall into the latter category.

The effects of spectacular failures of systems may be on such an immense scale as to warrant the label 'catastrophe'. Major accidents that result in large scale injuries, loss of life or material destruction provide dramatic examples: the cyclohexane explosion at Nypro's Flixborough plant (1974), the release of toxic substances into the atmosphere at Seveso (1976) and Bhopal (1986), the Zeebrugge ferry disaster (1987), the King's Cross fire on London Underground (1987), the Piper Alpha offshore rig explosion (1988), the Boeing 737 crash on the M1 motorway (1989), and many more. Failures in natural systems can be catastrophic, too: the Sahel drought and crop failures in the Horn of Africa leading to widespread famine, and the floods in the Sudan are examples in recent history. According to some scientists, the world's climate is getting warmer as a result of man-made atmospheric pollution and this could lead to a future catastrophe.

Economies are very complex systems and failures in them often affect many people. They involve socio-technical and other designed systems (e.g. the stock markets, industries, agriculture etc.), human activity systems (e.g. government policy making, trades unions, consumer activity etc.), natural systems (e.g. weather, pests etc.), and abstract systems (e.g. supply and demand formulae, planning models to reduce inflation, computer programs to simulate and predict effects of changes in exchange rates etc.). Depending on your point of view and purpose, you could choose to regard the economy as any of the four categories we have used. When a national economy fails, the effects can be catastrophic. The Stock Market Crash of 1929 and the worldwide slump that followed it still hold bitter memories for those who experienced hunger and deprivation as a result. Since the 1930s, the world economic system has been gradually modified so as to better withstand 'snowball' shocks.

Most failures do not occur in such an awe-inspiring public way.

Nevertheless, the consequences of 'lesser' failures can be serious for many individuals.

Exercises

2. In what ways could the failure of a business affect a large number of people?

7.3 Implications of systems failures

When a failure is detected, there is a great temptation to indulge in 'instant diagnosis'. The failure is attributed to a single cause on the basis of a quick reading of the situation. In some cases, experts can be given licence to do this on the basis of urgency and their own wealth of experience. For example, an accident investigator arriving at the scene of an explosion has to make some preliminary guesses as to the cause if only to ensure that there is no risk of further explosion.

However, non-experts and those assessing the situation from afar frequently do make pronouncements about the causes of failures that are little more than guesswork inspired by their own interests, values and prejudices. The two favourites trotted out as the cause of a particular accident are 'human error' and/or 'technical failure'.

Exercises

3. What is so unsatisfactory about attributing the cause of an accident solely to human error or to technical failure?

Our answer to Exercise 3 may have surprised you. After all, when someone makes a mistake and an accident occurs, surely that is 'human error'? If sprinkler systems do not operate during a fire, surely that is 'technical failure'? It is our contention that, in the main, human error and technical failure do not simply come out of nowhere as if they were acts of God. For example, inadvertence is a kind of human error to which everyone is subject, e.g. inadvertently putting your foot on the accelerator instead of the brake. However, many human errors occur as a result of receiving inadequate information, instruction and training. Latent defects do occur in engineered systems but many defects would be patently obvious to routine inspection and testing. Organisational, and especially managerial, functions such as selection, training, design engineering, maintenance and others need to be examined. Thus a *systems* approach to failures of whatever kind probes not only the technical and individual human aspects but also the *organisational* precursors to signs and symptoms of failure.

A systemic view of accident causation in particular is gaining popularity. Following two explosions at BP's Grangemouth plant in 1987 which resulted in fatalities and much damage, BP were fined £750 000. Mr Basil Butler, BP's Managing Director, later confessed to a British Institute of Management meeting in London: 'Incidents and accidents cost industry dearly in injuries, lives lost, shutdowns, and business interruption. It is now apparent (that) poor safety management can be as much to blame as human error or plant failure'.

7.4 The formal system model

Anticipation and prevention of failures requires understanding of the system concerned and how it works. Models (sometimes called paradigms) are useful tools for gaining understanding. There are countless possible models. A series of particularly useful models are described in Chapter 8.

By comparing various aspects of the failure situation with appropriate models, it is possible to discern whether discrepancies or agreements occur and whether these are significant. Such comparisons can be very illuminating.

The formal system model (FSM) is essential to any systems failure study of a situation in which human activity forms a part. The FSM should always be the first model used for comparison. It addresses the question: was there a system *at all* immediately before the failure in the situation concerned? Use of the FSM can also often point towards other models that would be useful for comparison. Figure 7.1 shows the FSM.

As Fig. 7.1 shows, the FSM requires the presence of the following components within the system:

- a control (or decision-making) sub-system
- one or more operational (or executive) sub-systems
- a performance monitoring sub-system.

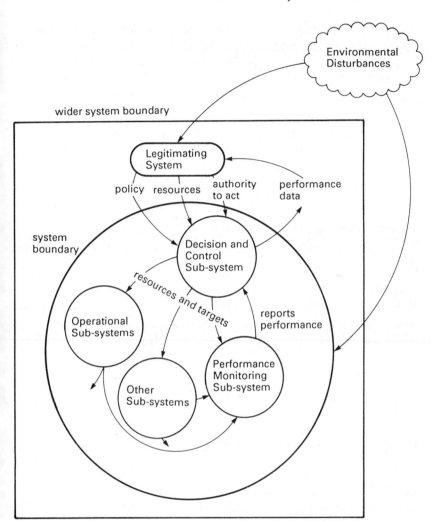

Fig. 7.1 The formal system model. (*source:* adapted from P. Checkland, *Systems Thinking, Systems Practice*, John Wiley & Sons, 1981; and L. Watson, *Systems Paradigms*, Open University T301 Course Material, 1984).

The control sub-system tells each operational sub-system what is expected of it and requires the performance monitoring sub-system to monitor the operational sub-systems and report back. For example, a Board of Directors (control) issues overall instructions and costs and revenue targets to Production, Warehousing, Marketing and Sales (operations) as appropriate. The Finance Department (monitoring) monitors the performance of departments in meeting those targets and feeds data to the Board. If targets are not being met, the Board has the opportunity to take control action to remedy the situation.

By comparing a particular situation with the FSM, discrepancies will be highlighted. Omissions such as no performance monitoring sub-system (a structural omission) or no control action (a process omission) are classic indicators of systems failures.

The FSM also requires the identification of a wider system that legitimates the existence of and activities of the system concerned. For example, the Board of a parent company might be regarded as the *wider* system of the example in the previous paragraph. The wider system:

- issues policy decisions and directives
- provides resources } to the control sub-system
- gives authority to act

The wider system also receives performance data from the control sub-system. You as analyst have to make a judgement in each case as to where the boundary lies between the system and its wider system. Both wider system and system are likely to be affected by environmental influences and disturbances.

Activity

For a situation that you know about that has been described as a failure, compare it with the FSM. Listing out the essential FSM features first and then checking their presence/absence/quality in the situation may be helpful. Then superimpose what you have onto the FSM diagram. Consider what understanding you gain about the system and its failure.

7.5 Summary

A failure exists in the mind of someone whose expectations have not been met. Many people can agree that a failure has occurred, especially in cases of accidents and misfortunes. Accidents and misfortunes may be regarded as emergent properties of systems that have failed. Comparison of the failure situation with various paradigms, and especially the formal system model, can illuminate the causes of failure and aid the prevention of future failure. Chapters 8, 13 and 14 continue the study of systems failures.

7.6 Suggested answers to exercises

1. It was a success in the sense that no one was injured, radioactivity was contained, and the plant safety systems operated as they were intended to. However, the massive disruption of surrounding communities, the chaotic management of the emergency and the consequent damage to public confidence would probably be described as failures by most people.

2. Employees lose their jobs, families experience stress, creditors and investors lose money, the local economy can be damaged because affected people spend less, and so on. The beneficiaries of a system are likely to be those who suffer most when the system fails. Of course, competitors are likely to gain as a result (i.e. perceive success).

3. Attributing the cause of an accident solely to human error or to technical failure suggests erroneous views of how accidents occur. First, it is well established that accidents have more than one cause and usually many causes. Second, although human error and technical failure may form part of a causal explanation, they are essentially *symptoms* of more fundamental causes rather than being *the* causes. In other words, human error and technical failure along with the accident event are emergent properties of a *system* that has failed.

Chapter 8
Using Systems Failures Ideas

Overview

There are a large number of models that can be used to analyse systems failures. Especially useful are control, communication, engineering reliability, and human factors models.

Control models concern the maintenance of a system in a desired state. Control processes may be either adaptive or non-adaptive, either linear or non-linear, and either continuous or discrete.

Communication models include human–human, human–machine, and machine–machine communications. System noise affecting encoding, communication channel and decoding is a common cause of failure with all three types.

Engineering reliability models concern failure mechanisms and failure prevention in engineered systems and their components. Such models include fault trees, cascades, common mode, and failure modes and effects.

Human factors models derive from the broad range of applied psychology and social sciences. Examples include stress, learning and group behaviour.

8.1 Introduction

Chapter 7 introduced the concept of system failure and the idea that studying a system failure leads to greater understanding of both the system and how and why it failed. Such knowledge can contribute to better system design and the prevention of future failures. In addition to the formal system model, there are many other models that you can apply to the failure situation, a selection of which are described in this chapter. Figure 8.1 depicts an organisation chart of some important models or paradigms.

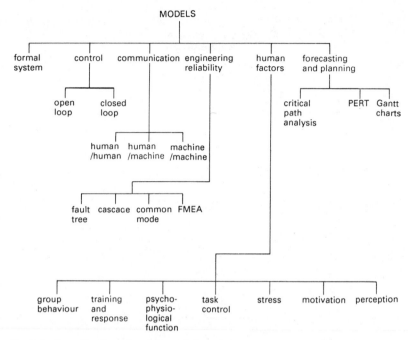

Fig. 8.1 Organisation chart of useful models.

8.2 The control model

As discussed in Chapter 7, the FSM is always the first paradigm to compare with your situation. This initial comparison may suggest a lack of control. Comparison with the control model should then provide greater detail of the inadequacy and confirm your initial findings. Even if FSM comparison does not indicate lack of control, a comparison with the control model is always worthwhile. Figure 8.2 depicts the control model in a simple form and Fig. 8.3 shows an application.

Control per se is an action or process which a system or sub-system applies to itself so as to maintain a desired state. Control processes may be differentiated by the frequency with which monitoring occurs – either continuous or as discrete samples at preset or irregular intervals. Control processes also differ according to how they function. Linear control processes are those in which control action is proportional to the discrepancy between the actual state (i.e. output) and the desired state (i.e. reference value). With non-linear control, however, no control action occurs until a pre-determined discrepancy is met.

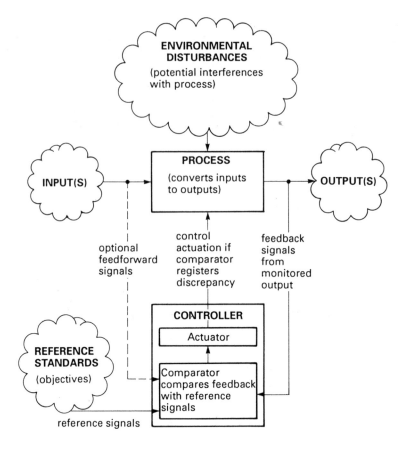

Fig. 8.2 The control model.

Exercises

1. Give some examples of (a) continuous monitoring, and (b) discrete or discontinuous monitoring.

2. Give some examples of (a) linear control, and (b) non-linear control.

Effective control requires certain conditions to be met:

- the process to be controlled must be understood
- inputs and outputs must be capable of being monitored reliably and at suitable frequency (e.g. with suitable measuring devices)

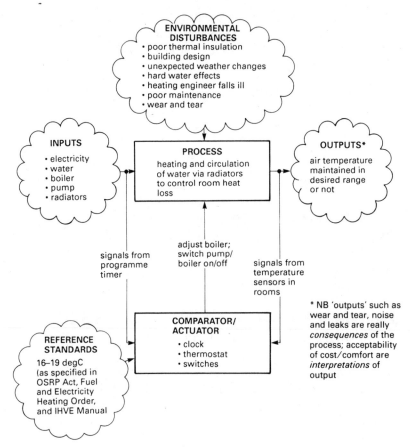

Fig. 8.3 The control model with the office central heating system superimposed to show both feedback (adaptive) and feedforward loops.

- there must be an adequate communication link between monitor and controller
- reference standards must be compatible with outputs being monitored
- time delays between control action and control effect should be within tolerable limits and there should be no overshoot

Exercises

3. The rate at which chemical reactions progress increases dispropor-
tionately with a rise in temperature, i.e. for every 10 deg C rise in

temperature the reaction rate roughly doubles. Some chemical reactions are prone to cause 'runaways', i.e. the reaction itself releases heat which if not dissipated speeds up the reaction which in turn releases even more heat and so on. If temperatures and pressures were monitored as outputs from such a chemical reactor by discrete sampling every 15 min, what problems could arise?

4. Give an example of technical overshoot.

5. Give an example of the kind of organisation that typically experiences time delays between control action and control effect.

8.3 Communication models

Communication failures are characteristic of failures in human activity systems but they may also occur in other types of system. There are three kinds of communication to consider: human–human, human–machine, and increasingly machine–machine. A frequently quoted example of communication failure between humans concerns the story of the telephone request from a front-line commander to his headquarters: 'Send reinforcements, we're going to advance' which the young corporal taking the call understood as 'Send three-and-fourpence, we're going to a dance'. Obviously, although the originator's message was received it was misinterpreted by the receiver. In a communication process, there are numerous opportunities for interference as shown in Fig. 8.4.

The two main kinds of interference are language difficulties and system noise. Obviously, communication becomes more difficult if the language of the receiver differs from that of the originator. Even people apparently using the same mother tongue can run into difficulty if one person is using special vocabulary that the other does not know. This latter phenomenon is sometimes called 'blinding with science'. Christopher Rowe (1985) in a case study of failure in computerised stock control quotes a company manager thus:

'I remember we were presented with this big orange folder, all about the computer we were going to get. I must confess it didn't mean much to me: it was all in computer jargon. But I didn't want to look daft so I didn't say anything. They explained it and it seemed reasonable, but I didn't really understand it. Of course it was decided by then that we were going ahead with it.'

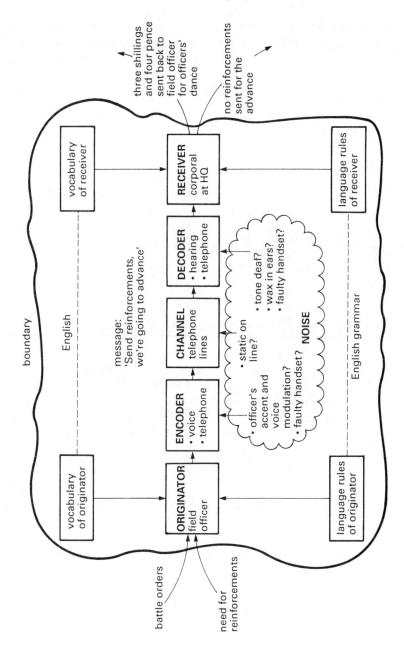

Fig. 8.4 The human-human communication model with example superimposed.

The author, during research into the introduction of computers, has been told by a number of experienced users about their communication difficulties with computer departments. Some have referred to computer staff as 'speaking computerese' (Fig. 8.5).

Most trades and professions use their own special language. Sometimes terms are identical with words used by the general public but with quite different meanings. For example:

term	specialism	meaning
gutter	printing	where facing left and right hand pages join (not the road/pavement junction)
liquor	brewing	water (not alcoholic spirit)
detect	policing	solving a crime (not uncovering it)
hard problem	systems	structured and quantifiable (not necessarily a difficult problem)

No doubt you can add your own examples and recall or imagine misunderstandings involving them. Newspaper headlines are renowned for semantic puns such as 'Police found safe under blanket'. Did the police find a safe under a blanket, or were some policemen found unharmed under a blanket?

Misunderstandings of vocabulary can have important safety implications; an accident may be considered to be a system failure and quite often communication failures are involved. For example, chemicals may have similar looking and sounding names but quite different chemical and toxicological properties. Benzene and benzine are pronounced the same but are quite different. Nitrogen is referred to by chemists as an inert gas which may imply (wrongly) to non-chemists that it is harmless. Photographer's hypo (sodium thiosulphate) is quite different from the hypo (sodium hypochlorite or bleach) as understood in some chemical works. Ice (frozen water) is quite different from dry ice (frozen carbon dioxide).

The term 'on-line' is used widely in computing, manufacturing and the process industries to mean connected and powered up for processing. When disconnected or shut down, processes are said to be 'off-line'. However, some chemical processes that are officially

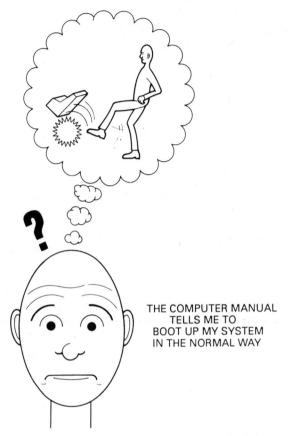

Fig. 8.5 The technical language of specialists can cause communication problems.

complete and off-line may contain residual reactants which may continue to react and lead to a runaway as described in Exercise 3. If workers understand off-line to mean that temperature monitoring is no longer required, then a dangerous situation could rapidly escalate unchecked.

All the above examples are taken from case histories where accidents have occurred through such communication failures.

Exercises

6. The standing safety instructions in a factory stated that 'all sealed radioactive sources must be checked by the Authorised Person each working day'. How could this result in a communication failure?

Referring to Fig. 8.4, encoding includes such things as voice, pen, and keyboard. Channels include the post, telephone, electronic mail, etc. Decoding refers to the receiver's processes of vision and hearing. With written communications, encoding and decoding are more structured and formal than with verbal communication. In the communication context, noise refers not just to unwanted sound but to *anything* that interferes with or stops communication.

Exercises

7. What are the advantages and disadvantages of written communications?

8. Give examples of communication noise.

Encoding and decoding are also features of the human–machine communication model depicted in Fig. 8.6. Noise can interfere. For example, in the Three Mile Island nuclear power station accident an indicator light in the control room was masked by a caution tag. When a temperature indicator read 285 deg F the control room operators misread it as 235 deg F because that is what they expected it to be.

However, not only do you need to consider noise that interferes with the human sub-system but also noise that affects the machine. For example, machine noise may arise from:

- failure to validate data input (e.g. computer data)
- failure to provide internal error checks (e.g. in system software)
- damage to remote sensors causing false display
- corruption of signals
- component wear or failures.

Communication between machines usually entails electronic signals passing between two or more computers or machines that are computer-controlled. The distinctions between computer technology and telecommunications is rapidly blurring as more and more computers send data to each other via the telecommunications network and telephone products are given features of computer terminals. The possibilities of communication failure as a result of noise are as described in the previous section. Some of the more common ones are component failures, faults at exchanges and channel interruptions such as telephone lines cut or shorting out.

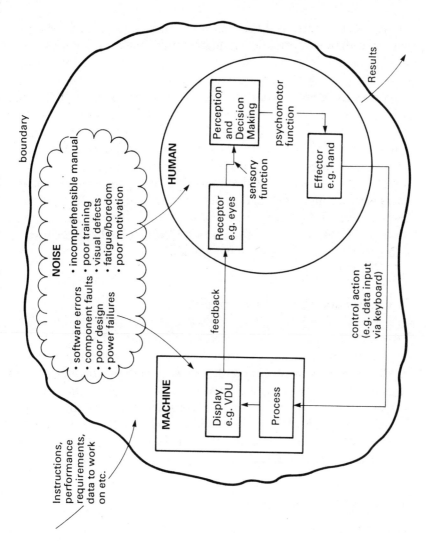

Fig. 8.6 The human-machine communication model with example superimposed.

Another source of failure for machine–machine communication is information overload, which also can occur with human–human and human–machine communication. When this occurs, the originator is encoding and sending more data more rapidly than the receiver can cope with. An evocative example of this was provided by the so-called 'Big Bang' in the City of London in 1986 when the Stock Exchange was de-regulated and an on-line computerised information system called

SEAQ was introduced. Previously, brokers and traders relied on various different information systems to guide their transactions which were done mainly over the telephone. SEAQ provides a much faster monitoring of deals and thus requires much faster decisions and responses from users. Transactions are intended to be made through SEAQ rather than on the telephone. As SEAQ is an integrated system aimed at replacing several others, the number of users at any one time is high. In order to make SEAQ easy to use, SEAQ data are fed to the TOPIC computer which displays SEAQ data in Viewdata format on users' terminals. In the early days of Big Bang, sometimes so many users were on-line that SEAQ was overloaded and TOPIC could not keep up with SEAQ. Users were unable to rely on share data displayed on their TOPIC screens. It should be added that lessons were learned and improvements have since been made.

8.4 Engineering reliability models

Engineering reliability is concerned to ensure that there is an acceptable period of time before failure occurs. Engineering reliability models relate to engineered systems. Although you may not be an engineer, if you are in any way involved in the design, manufacture or supply of products you need to consider this section carefully. Legislation designed to protect consumers and users of products, coupled with increases in product liability claims, will require a greater emphasis on system reliability.

Measures of reliability

The term reliability in an engineering sense refers to the degree of confidence that can be placed in a component (or system of components) fulfilling its duties in service. In other words, reliability is more concerned with success than with failure. Engineering reliability is a numerical concept and is usually quoted as a probability value in the range 0.0 to 1.0. A reliability of 1.0 means that it is expected never to fail in service whereas a reliability of 0 predicts certain failure. The period of time over which the component or system is expected to operate is called the mission time.

Exercises

9. If an insurance engineer carried out a thorough six-monthly examination of a powered lift and stated in his report that the overrun device had a reliability of 0.4, what would you conclude from this?

Advanced panel

Component failures do not occur at a uniform rate. For example, if a batch or sample of new components were tested under service conditions, there would be a relatively high number of failures early in the test period. These systematic failures are usually due to poor manufacturing methods and lack of quality control. This high initial rate is followed by a rapid falling off to a constant failure rate for a long period. At the end of this stable period, i.e. useful life, the failure rate rises again as components wear out. Plotting failure rate against time gives the well-known 'bathtub' curve as in Fig. 8.7.

It is possible to estimate the number of failures likely to occur in a component's service life from the formula $R = e^{-wt}$ where e = the exponent (2.718), ω = constant failure rate (i.e. the flat part of the bathtub curve), and t = operating time in hours. For example, an electric pump has a stable failure rate of 0.001 per hour and operates for a

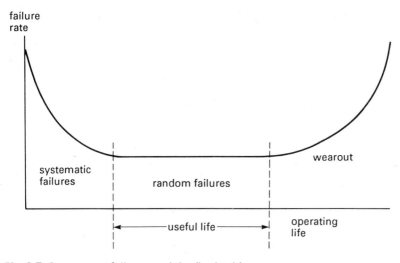

Fig. 8.7 Component failures and the 'bathtub' curve.

period of 18 hours. Therefore, $\omega t = 0.001 \times 18 = 0.018$. Referal to published exponential tables gives a value of 0.98 or 98% reliability. Thus, if 100 similar pumps operated for 18 hours each, two of them would be expected to fail.

Another measure of reliability is the mean time between failure (mtbf) which is an estimate of component life and is usually derived from testing a sample of components to destruction. A safety factor can then be applied, with the intention that in service a wide enough safety margin will exist between operating conditions and failure conditions. For example, the maximum service load might be chosen to be one quarter of the destructive load found in a laboratory test. This gives a safety factor of 4 and, in theory, a four-fold safety margin. Thus, a maximum safe working load of 2.5 tons might be set based on the assumption that failure will occur at 10 tons. Similarly, a maximum service life of 5 years might be set, based on an mtbf of 20 years.

Exercises

10. On what main assumption do safety factors rely? What adverse consequences of this assumption are there?

Since a system comprises interconnected components, it is easy to see that component failure may lead to system failure. In other words, system reliability is a function of component reliability. However, although each component's reliability may be regarded as constant the way they are interconnected may have a profound effect on system reliability. Also, failure of one component may accelerate the failure of another, e.g. fan belt failure may lead to radiator failure.

Advanced panel

Figure 8.8a shows the components of an alarm system connected in series. It is obviously a vulnerable system because failure of any one component will result in system failure. Figure 8.8b shows the system expressed as a reliability block diagram. The reliability of this series system is calculated by multiplying all the component reliabilities together:

$R(\text{system}) = R(\text{sensor}) \times R(\text{wiring}) \times R(\text{control}) \times R(\text{alarm})$

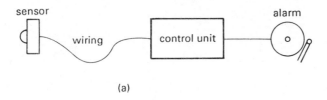

(a)

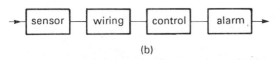

(b)

Fig. 8.8 Reliability of components in series. (a) Schematic diagram of series alarm system. (b) Reliability block diagram of series alarm system. (Adapted from: R.T. Booth, H. Raafat and A.E. Waring, *ST2 Machinery and Plant Integrity*, Occupational Health and Safety Open Learning, 1988).

Exercises

11. You are designing an alarm system from specified components with reliabilities R(sensor) = 0.90, R(wiring) = 0.97, R(control) = 0.95, and R(alarm) = 0.94. You feel that you can improve the system reliability considerably by selecting a different sensor with a reliability of 0.95 and guess that the system reliability will now be about 0.95. Will you achieve this improvement?

In order to increase system reliability and reduce the effects of component failure on the system, it is common practice (especially in electronics) to introduce redundancy, i.e. duplication or triplication of critical components. In space craft, for example, it is usual to have three identical computers running in parallel. The idea behind redundancy is that component failures are independent and should one fail then one or more redundant partners are available to take over its function. In parallel active redundancy, all components operate under full load. Figure 8.9a shows the previous alarm system with redundant sensors and Fig. 8.9b is its reliability block diagram.

If the sensor reliability is 0.90 as before, the reliability of the duplicate sensors is given by:

R(dupe. sensors) = $1 - (1 - 0.90)(1 - 0.90)$
$\qquad\qquad\qquad = 0.99$

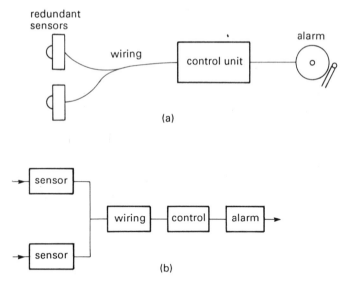

Fig. 8.9 Reliability of system with parallel active redundancy. (a) Schematic diagram of alarm system with redundant sensors. (b) Reliability block diagram with redundant components. (Adapted from: R.T. Booth, H. Raafat and A.E. Waring, ST2 Machinery and Plant Integrity, Occupational Health and Safety Open Learning, 1988).

Putting this value into the formula for series system reliability as before:

R(system) $= 0.99 \times 0.97 \times 0.95 \times 0.94$
$= 0.86$

This represents an increase of 8% in system reliability over the previous value of 0.78 (see answer to exercise 11) but still nowhere near good enough for a practical alarm system. Individual component reliabilities, even with redundancy, would have to be better than the hypothetical values used in the above examples. If you are interested in learning more about reliability in safety critical systems, we recommend you to read *Machinery and Plant Integrity* (1988) by Richard Booth, Hani Raafat and Alan Waring (see Useful Reading).

Outcomes of component failure

A component failure may or may not cause a system to fail. Engineering design seeks to ensure that component failure does not cause complete system failure. There are four kinds of outcome of component failure:

- *fail-safe*: the failure usually results in the orderly shut-down of the system which then cannot be restarted until the failed component has been replaced. In the case of an alarm system whose prime function is to alert people to danger, the system may be designed to sound the alarm or a special alarm signal in the event of a component failure. NB 'fail-safe' does *not* guarantee 100% safety.
- *fail-soft*: by design, the failure results in the orderly shut-down of the system which then can only be restarted by human intervention. Typical are automatic cut-outs and trips that operate when components are temporarily overloaded. In many systems, such as machinery having dangerous parts that people may come close to, a fail-soft design cannot be justified.
- *fail-run*: the system continues to run in spite of component failure and is deemed safe to do so because of its design, e.g. redundancy techniques.
- *fail-to-danger*: here the component failure definitely causes a dangerous condition. The danger may be within the system itself for example if the power supply to a badly designed interlock guard on a power press fails the dangerous parts of the machine may still be contactable. Alternatively, the danger may arise because an alarm system, for example, is no longer capable of raising an alarm. When components fail in systems designed to fail-to-danger, often the failure remains unannounced and may only be detected by maintenance checks or when an accident occurs.

Technical failure categories

Component failures often occur because of discrepancies between expected and actual conditions. Typically, such discrepancies fall into the following categories:

- the loading and/or environmental conditions in service are more severe than those assumed in the design specification
- the component or structure is not made to the design drawings or specification
- the design itself is inadequate for even the expected forces and environmental conditions in service
- maintenance and inspection procedures are faulty or do not occur.

Figure 8.10 summarises these categories. There are numerous examples of catastrophic failures where some or all of these categories were involved:

Forces/environmental conditions

		Expected	Unexpected
Design	Adequate	Very low probability of failure	Low probability of failure
	Inadequate	Failure probable	High probability of failure

(a)

Manufacture

		To design	Not to design
Design	Adequate for service	Low probability of failure	Uncertainty so assume high probability of failure
	Inadequate for service	High probability of failure	Uncertainty so assume high probability of failure

(b)

Inspection and maintenance

		Adequate	Faulty
Failure	Deterioration detectable	Low probability of final failure	High probability of final failure
	No deterioration	Low probability of final failure	High probability of failure as onset will not be detected

(c)

Fig. 8.10 Tabulation of technical failure categories (a) Designing for forces in service. (b) Manufacturing from design. (c) Inspection and maintenance in service. (*Source:* R.T. Booth, H. Raafat and A.E. Waring, *ST2 Machinery and Plant Integrity*, Occupational Health & Safety Open Learning, 1988.)

- Brent Cross crane failure 1964 (overload; design specification for critical component not followed)
- Appleby-Frodingham steel works explosion 1977 (steel blanking plugs used instead of brass which led to corrosion)
- Littlebrook D hoist failure 1978 (overload; corrosion; poor state of repair; inadequate inspection; design faults)

- Flixborough Nypro explosion 1974 (temporary bypass design did not follow relevant British Standard and was inadequate for stress conditions; inadequate testing)
- Alexander Kielland oil rig collapse 1980 (service conditions harsher than design allowed for; design did not account for interdependence of components; inadequate inspection during construction and in service)
- Ferrybridge cooling towers collapse c. 1961 (design did not account for gust velocity due to Bernoulli effects between the front towers; wind hit rear towers at very high speeds).

In order to prevent such disasters, design and construction practice need to be based on procedures whereby:

- a set of standards of required performance is laid down
- failure characteristics of components are estimated
- the effect of individual component failure on other components and on the overall system is estimated
- levels of performance are checked against the required standards

Four models are particularly useful: fault trees, cascade model, common cause or common mode paradigm, and failure modes and effects analysis (FMEA).

Fault trees

Fault trees chart the conditions necessary for a particular failure. Although they can be used for failure of purely engineering components and systems, we feel that they should also incorporate soft elements where necessary. We said in Chapter 7 that technical faults are symptomatic of human and organisational failures.

Fault trees enable prediction of how a failure might occur. Each tree is built up as a series of levels or hierarchies as in an organisation chart. Levels are linked by logic gates as in Fig. 8.11. Starting with the failure event (fire), the three conditions needed are a source of combustible material AND a source of oxygen AND a source of heat. Sources of combustible material are either gas OR flammable liquid OR other organic material. In order to have sufficient heat for combustion, there must be both a source of ignition AND a high enough temperature. You carry on building your tree downwards in this way until your purpose is satisfied.

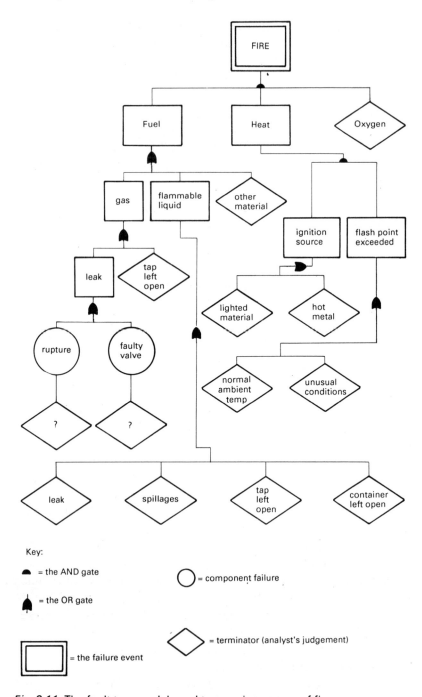

Fig. 8.11 The fault tree model used to examine causes of fire.

Activity

As Transport Manager for a road haulage company you are concerned at the frequency with which a particular lorry is failing to start in the mornings. Two possible main causes come to mind: fault with the vehicle or the new driver is using the wrong procedures but you have not yet thought further about it. Construct a fault tree showing the possible causes of the failure.

Cascade model

In plain English, the cascade may be expressed as one event leading to another which leads to another and so on. Such a chain of failures is called the domino effect. However, in its simple domino form the cascade paradigm is not very useful because so few failures occur in such a simple linear way.

A much more useful cascade model is one in which one or more failures in the chain cause multiple and looping failures – the chain reaction. You have already met this paradigm in a different guise – the self-enhancing 'snowball' as depicted in positive loop causal diagrams (see Chapters 2 and 3 for examples). Earlier in this chapter, we described the runaway chemical reaction – a typical cascade.

Fault trees are useful for working out the failure components, but they cannot show the presence of a self-enhancing chain reaction. For this you will need to construct causal loop cascade diagrams. Chapter 14 shows some good examples of fault tree and cascade diagrams applied to the structural failure of system-built housing units.

Common mode paradigm

Common mode or common cause failures occur when all components of a specified kind have a common fault. Systematic faults like this are usually a result of poor design or manufacturing faults that are not identified by quality control. Typical examples are whole batches or models of car having to be recalled because the manufacturer has discovered a design or manufacturing fault. Sometimes common mode faults are only discovered as a result of similar accidents occurring. A recent example is provided by apparent common mode wiring faults found in a number of Boeing jets following the Boeing 737 crash on the M1 motorway early in 1989, although at the time of going to press the official investigation has yet to confirm the causes of the accident.

Exercises

12. Earlier we noted that parallel redundancy was an effective way to increase system reliability. How is reliability affected in such a system if redundant components possess common mode faults?

Failure modes and effects

Rather than showing *how* failures might occur which is the main use of fault trees and cascades, FMEA attempts to predict *what* failures might occur and what their effects might be. FMEA provides a systematic method for examining each component in a system, its function(s), what kind of failures it might experience, how such failures might operate, what effects the failures could have on the system, how the failures are or could be detected, and so on. The starting point is an inventory of all components which are then tabulated and examined as indicated above.

Typical applications of FMEA are in the detailed planning, design, construction, commissioning and in-service monitoring of safety critical systems such as nuclear power plant, chemical and petrochemical plant, aircraft, and spacecraft.

8.5 Human factors models

Human factors models are soft in character and encompass a wide range of models concerning psychology of the individual, group pyschology, organisational behaviour, sociology, social psychology, ergonomics, and training. Such paradigms are of particular interest to people engaged in human resource development (personnel, selection, manpower planning, training, etc.).

As there are too many human factors models to consider in this book, we have selected two as examples. One is the human–machine interface model, an ergonomics paradigm which you have already met in Fig. 8.6 in connection with human–machine communication. The other is the stress model as shown in Fig. 8.12. The stress model shows a person's needs and capabilities system which is separated from the external influences of the physical environment (temperature, light, etc.) and social environment (friends, enemies, human contact etc.). Where there is imbalance, stress and strain result.

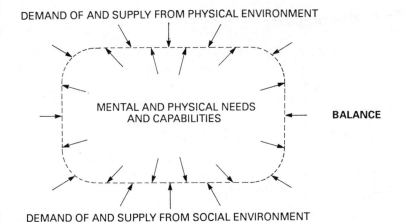

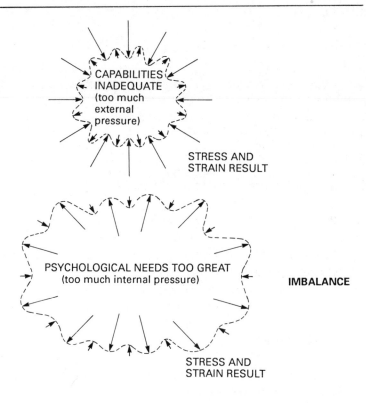

Fig. 8.12 The human stress model. (*Source:* A.E. Waring, OTA5 Office Environment, Office Technology Open Learning, South Bank Polytechnic, 1986.)

8.6 Analysing system failures

You now have at your disposal a large repertoire of models for comparison with an apparent failure situation. However, such comparisons will not be very effective or useful if made in an undisciplined way. What is needed is a systems framework for studying failures into which paradigm comparisons fit. A surgeon, by analogy, needs to be able to establish what systems and organs are showing signs of failure before deciding what the most appropriate instruments are. A framework for analysing systems failures is described in Chapter 13 with case studies in Chapter 14.

8.7 Summary

Models or paradigms are 'viewing instruments' that enable you to understand how particular aspects of a situation might have arisen or might operate. Some models are desirable (e.g. FSM, control and communication) whereas others are undesirable (e.g. fault tree, cascade and common mode). Both desirable and undesirable models can be used to illuminate systems failures.

8.8 Suggested answers to exercises

1. (a) *Continuous monitoring*: temperature control of car engines; temperature and pressure control in chemical plant; homeostatic (self-maintaining) control of body functions such as blood pressure, blood CO_2 etc.
   (b) *Discrete monitoring*: some gas monitors sample only every 20–30 seconds in order to prolong sensor life; human beings in plant control rooms may have to monitor series of indicators in sequence; stock audits may be performed, perhaps, once a week.

2. (a) *Linear control*: a driver steering a car; a driver using the footbrake to decrease speed.
   (b) *Non-linear control*: in a refrigerator the pump only switches on when the temperature inside the 'fridge rises to a particular value; companies chase unpaid invoices when the credit period is up; companies chase bad debts once they exceed a minimum amount.

3. The reaction may remain stable for a long time, but if it should become unstable and a runaway occurs the sampling frequency may be too low to identify a rapid and probably dangerous rise in temperature and pressure. Even with continuous monitoring, the communication between monitor and controller must operate at least as fast as conditions are changing. This did not happen at the Three Mile Island LOCA (see Chapter 7) where a computer printer was printing vital data on which control staff had to act but more slowly than signals to the computer were changing.

4. *Technical overshoot*: when steering a boat, turning the rudder in one direction too sharply causes the boat to overshoot the intended new course; prediction of overshoot allows counter-control action (turning the rudder in the opposite direction) to reduce the overshoot.

5. *Organisational control*: large bureaucratic organisations such as public services and higher education establishments often take a long time to institute major changes owing to the inertia such bodies exhibit in their administrative processes.

6. The phrase 'each working day' is open to several interpretations. In the factory in question, workers understood it to mean each standard work day (i.e. Monday to Friday) whereas the factory operated seven days a week. The 'working day' meant each day the radioactive source was in use.

7. *Advantages of written communication*: formality of grammar reduces ambiguity; larger written vocabulary aids precision of meaning; preparation reduces ambiguities and errors; message is more permanent.
   Disadvantages: receiver cannot interrupt or query for clarification; receiver less able to gauge originator's intentions as no intonation, body language etc; message may get lost or misdirected.

8. *Communication noise*: extraneous sounds, false assumptions of the originator about the knowledge of the receiver; poor expression of ideas by the originator; the receiver's false expectations of message content based on past experience; the receiver's visual or hearing difficulties, and so on.

9. A reliability of 0.4 is equivalent to only a 40% chance of success. Therefore, the overrun device is more likely to fail than to successfully complete its mission, i.e. work properly for the next six months.

10. The *safety factor approach* assumes that test data from a laboratory accurately reflect how the component would actually behave in a wide variety of service conditions. Service conditions may be far more rigorous than allowed for. The safety margin calculated in this way thus may be far wider than is warranted and so may create a false sense of safety.

11. *System reliability* with the original components is 0.78. With an improved sensor, system reliability increases to 0.82; an improvement but nowhere near what you expected. Overall reliability can never be greater than that of the least reliable component. Try some different component reliability values to satisfy yourself that this is so. Consider the analogy: a chain is only as strong as its weakest link.

12. Parallel redundancy itself cannot overcome *common mode faults*. A technique that limits the effects of common mode faults is called diversity – using different components and modular designs. For example, instead of using a common power supply, which if it failed would stop a whole system which depended on it, it might be possible to power critical sub-systems independently by, say, batteries. Hospitals and other premises use stand-by generators that automatically provide power in the event of the mains supply failing.

PART 2

A Closer Look at Using Systems Ideas

Part 2
A Closer Look at Using Systems Ideas

In Part 2, it is assumed that you are fully conversant with the ideas, principles and techniques described in Part 1.

Part 2 develops hard systems analysis, soft systems analysis and analysis of systems failures with step-by-step guidance on how to carry them out. Although a step-by-step approach is needed in order to describe what to do, you always need to be wary of getting into the mental rut of 'recipe thinking'.

Chapter 9
Hard Systems Analysis

9.1 Introduction

In Chapters 3 and 4 of Part 1, you were introduced to the concept of hard systems and to various examples relevant to work. We also gave a very brief introduction to the hard systems approach to management problem solving. In this and the next chapter, we expand upon hard systems analysis. First, however, you need to review hard systems thinking in relation to decision support.

9.2 Hard systems revisited

Some problems may be solved satisfactorily by using past experience or practice as a guide. In fact, most day-to-day problems fall into this category. If the car will not start, for example, a few obvious possible causes can be quickly checked. If you are lucky, you find the cause, deal with it and proceed on your journey. If you fail to find the cause, you can still solve the problem of a delayed journey by catching a taxi, bus, train or even by walking. A 'quick fix' is the appropriate response to the situation until you or the garage has time to find the fault and rectify it.

With problems that involve higher stakes, i.e. the costs of failing to achieve goals are high, a more systemic (or at least systematic) approach to solving them is appropriate. When such problems are well understood and the range of possible solutions have been tried and tested over a long period, a 'quick fix' on a more elaborate scale may be quite satisfactory. Here, a formal problem-solving procedure should be considered as outlined in Fig. 4.1 of Chapter 4. The nature of the problem and the range of solutions are considered by the problem solver to be self-evident; in principle, all that is required is a systematic appraisal and selection procedure. Figure 2.12 in Chapter 2 gives an example of an application to the problem of deciding what should go into a product promotional campaign. Formal problem solving like this would probably be appropriate in a market that was mature and stable and one in which the product manufacturer had considerable experience. Nevertheless, quality of experience is not always related to length of experience and so cannot be relied upon implicitly!

Formal problem solving, however, is likely to be effective only in cases where uncertainty about the problem and possible solutions (i.e. about cause and effect) is minimal, the problem setting is stable, and the level of complexity is low. In cases where these conditions do not apply but nonetheless the problem appears to be well structured and quantifiable, a hard systems approach is likely to be more effective. The analyst must decide when formal problem solving will suffice and when to use hard systems analysis.

The choice of an inadequate problem-solving tool will reveal itself only too well when the solution has been implemented. The solution may not work at all, or may work counter-intuitively and make the problem worse, or may produce unexpected new problems.

Exercises

1. Bearing in mind the full definition of a system, what do these kinds of failure of solutions suggest about the problem solver's understanding of the nature of the problem?

Complexity is often, but not always, a function of size and scale. Typical are problems of public transport. British Rail, for example, has often been criticised for running overcrowded trains. In order to make travel more comfortable for commuters coming into London, British Rail increased the number of carriages on an early morning train which previously had been very overcrowded each day. Within a short time, however, the extended train also became very overcrowded each day. The new overcrowding arose because some commuters who had previously switched to an earlier train to avoid overcrowding switched back to their preferred train after it had been extended. The result of British Rail's solution was a continuation of overcrowding in the 'problem' train and an earlier train now running half empty.

Problem solving that tinkers with the system of which the problem is an emergent property often results in 'migration' of the problem. Those of you who regularly experience traffic jams at particular places and times will recognise how attempts to solve queueing problems at one location often simply shift the queue further up the road. For example, main arterial roads into London such as the A40 have undergone major alterations in order to prevent existing queueing problems at the many roundabouts. The massive investment of capital certainly reduced queueing along the A40, but the net effect has been to shunt all the local queues into a single queue nearer London.

One of the contributors who is engaged in road traffic research has identified a similar problem with road traffic accidents (see Boyle and Wright 1984). When accident black spots (i.e. sites on the road traffic network that have a relatively high number of accidents) are treated to reduce accidents, for example are provided with anti-skid surfacing, pedestrian refuges, increased lighting etc., the number of accidents at such locations goes down. However, the number of accidents at untreated locations in the vicinity shows a corresponding increase. The suggestion is that drivers show increased attention at treated blackspots and surrounding areas and then relax their attention elsewhere. The accident problem has migrated rather than been solved.

Problem solving represents an intervention into a system. A further graphic example of problem solving that was not sufficiently systemic in its approach is provided by the World Health Organisation's intervention into a public health problem in Borneo. Traditionally, Borneo Dayaks live in communal long huts, often with several hundred inhabitants. Standards of health among such people were low and insect-borne diseases were identified as a prime cause. The WHO solution was to systematically spray all such dwellings with DDT. A dramatic improvement in the inhabitants' health followed, as intended and expected. However, a cascade of wholly unexpected problems then followed as indicated in Fig. 9.1.

The DDT entered the food chain of insects and mammals who were natural cohabitants in the huts. DDT-contaminated cockroaches were eaten by lizards which in turn were eaten by cats. At each successive stage in the food chain, the DDT dose became more concentrated to the extent that the cats received fatal doses. Soon the villages were invaded by a plague of woodland rats that the cats previously had kept under control. These rats brought fleas and other parasites with them which were infected with sylvatic plague. Eventually, the RAF had to

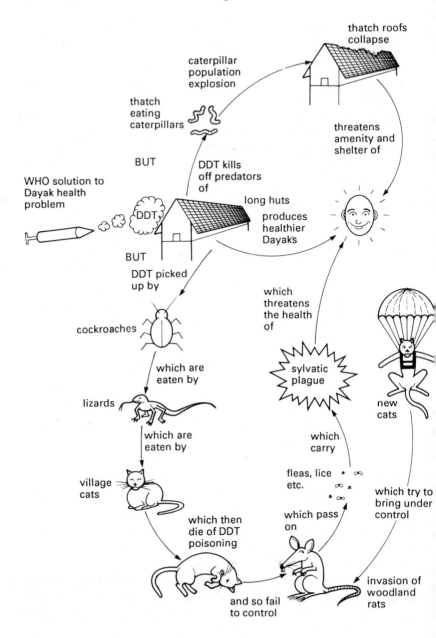

Fig. 9.1 Example of counter-intuitive outcomes to problem solving.

air-drop new cats into isolated villages to bring the rats under control again. If the sylvatic plague problem were not enough, the DDT had also killed off the natural predators of caterpillars that fed off and caused minor damage to the thatch roofs in the villages. Soon a plague of caterpillars was devouring the roofs and causing them to collapse (see Holling and Goldberg 1973).

The use of DDT in this case represented an obvious solution to an 'obvious' problem – a quick fix to what was *thought* to be a well-known and understood problem. However, as is clear from the above, the health problems of the long hut people were much more complex and were symptomatic of a finely balanced natural system. The people were part of an ecosystem which the DDT blockbuster solution threw into disarray.

A more recent example is the disarray caused to a finely tuned agricultural system on the island of Bali. In 1983, Western aid agencies pressed Balinese farmers to adopt more 'efficient' methods of rice production and to aim for rice export as part of the 'green revolution' in SE Asia. The Balinese had developed over a thousand years an intricate system of waterways and irrigation channels feeding down from volcanic Lake Batur. Under the new scheme, new dams and canals were built, new cropping patterns were started and new 'miracle' rice varieties were introduced. The expected economic boom did not occur.

Wherever the miracle rice was introduced new 'miracle' pests arrived. Use of insecticides resulted in a decline in soil fertility. According to studies by Drs Kremer and Lancing from the University of Southern California, the intervention failed because foreign experts had relied on Western assumptions about technological 'cures' and had not studied the social and ecological dimensions of the Balinese agricultural system. It became clear that pests were kept to a minimum by planting all the rice at the same time and cropping it at the same time. That required a lot of coordination, a fact that had gone unnoticed.

Water sharing down the waterway network from Lake Batur had to be at an optimal level so that no farmer went short. The coordination needed for this also went unnoticed by aid workers. It transpired that what was holding all this cooperation and coordination together was a system of elected farmer representatives who worked closely with priests at a series of strategically placed water temples on the waterway network. Water sharing rituals were important to the farming culture and were symbolic of the social cooperation needed for successful rice growing, i.e. the hierarchy of dependence on upstream neighbours.

Systematic procedures such as formal problem solving are inadequate for tackling real-world problems where complex relationships exist and simplistic assumptions about them are unwarranted. The hard systems approach tries to avoid the pitfalls of oversimplifying the problem by adopting a series of steps designed to develop a wider, systemic perspective and an awareness of all the significant aspects of the problem situation. The method is outlined in Chapter 4 and Fig. 4.3 summarises the steps involved. Refer back to them if you need to. We are now going to run through the method step by step in some detail. Remember, however, to try to avoid seeing any systems method as if it were a simple recipe for success.

9.3 Step 1: Groundwork

The groundwork involves identifying, and establishing a working relationship with, the client set, i.e. all those with whom the study seeks to gain credibility. This process may seem easier than it sometimes is in practice. If you are an external consultant, for example, you have to negotiate access to the client set. The person who commissions you to do the work (your effective client) is usually only one of a number of people in the organisation who have an interest in the study and its outcome. Establishing what those various interests are (who owns the system concerned, who owns the problem, who can give or withhold approval, what attitudes are prevalent? etc.) often takes time and skill. For example, while you are trying to extract information from key people, *they* will be appraising *you* and may well be guarded in their responses. Inexperienced analysts often fail to recognise that access to and relationships with the client set have to be negotiated *continually* throughout the study and not as a once-off step at the beginning. Similarly, awareness of the client set's Weltanschauung is important throughout. In addition to any written contract, there are bound to develop informal contracts between you and individual members of the client set. For example, you may be able to furnish the name of a useful contact who may help one of the client set with a problem quite unrelated to the one at issue. He or she may be able to facilitate an introduction to a previously unidentified key member of the client set. Building and maintaining trust and confidence is essential for the study to be effective.

Even if you are an 'insider', perhaps within the department with the problem or from another part of the organisation, you will still have to negotiate and maintain access. Defining or explaining exactly what needs to be done is not easy. Since a hard systems study is going to form the basis of the contract between the client set and you as problem-solving facilitator, it is clearly worth taking time to get it right. If you are also the problem owner and/or system owner, formally clarifying the nature of the problem and the problem-solving task is still necessary.

There are three practical ways to clarify these matters. First, is the fairly obvious step of agreeing formally with the client set what the project topic is and its likely scope. This will at least avoid a response to the first progress report that is either shocked silence or utterances such as 'typical consultants – never do what they're asked to do'. You can also ascertain whether your client is in a position to act on your recommendations or whether he or she is seeking a convincing argument to persuade colleagues that action is needed. In the latter case, the problem may need restating.

The second practical thing to do is to find out the client set's world-view in relation to the task and whether there is general agreement about the nature of the current position. Is the topic regarded as a problem or as an opportunity? Is the client set composed of risk takers or risk avoiders? This question is not only relevant to how possible solutions are presented but will also be a constraint on the avenues to be explored for potential solutions. What does the client set or organisation as a whole view as being acceptable action? Attitudes towards research and development, loyalty to employees, standing in the community, dealings with other organisations and so on will clearly be of major importance throughout the project.

Exercises

2. If you discovered markedly different values among the client set, clashes of opinion on major relevant issues, personality clashes or other signs of conflict, how ought you to proceed?

Third, find out as early as possible what the client set would consider to be a successful outcome. Establishing clear and realistic expectations is both a way of refining the topic to be addressed and further cementing the contract between client set and you as problem solver.

Very often, however, the client set does not have a clear idea of what is required and the project brief may be couched in very general terms. Equally, very large projects are rarely agreed upon without a well-structured project proposal from the consultant which could be on a competitive tender basis. In such cases, you would be briefed by the client and would have to marshal and assess all the relevant information right at the beginning of the groundwork stage in order to formulate a project proposal.

Once a clearer picture of the problem situation has been established, you can define your own commitment to the project in terms of the problem, your role, the problem owner's main objectives and your main objectives.

9.4 Step 2: Awareness and understanding

This step may be implicit in other problem-solving approaches. The hard systems approach makes it explicit and in particular encourages the use of systems description. The discipline of examining the system which contains the problem and identifying components in systems terminology should enable you to gain a clear perspective of where the problem fits and what the possible effects of trying to solve the problem will be on the rest of the system. The aim is to reduce the number of counter-intuitive outcomes of any change or at least to anticipate those outcomes.

A structured approach to systems description at this pre-analysis stage is that outlined in Chapter 2. Refer back if you need to. The diagramming techniques learned from Part 1 of this book can be used to advantage here. No one diagram type is inherently better than any other at this stage, although one would normally expect to see at least a systems map to set the overall scene and a diagram showing processes within the system of interest. As a prelude to system mapping and drawing influence diagrams, spider graphs or spray diagrams can be very useful as they may reveal a number of relevant systems. Teasing out several, perhaps overlapping, systems may enable the problem to be considered from different perspectives.

You have now set the scene and selected a system that seems to hold the key to a fruitful resolution of the problem. Detailed analysis of what goes on in that system must now be made. It is this detailed analysis which shows up what the knock-on effects of change are likely to be.

Influence diagrams and causal loop diagrams can be useful at this stage. Three mini-case studies serve to demonstrate how to establish patterns of influence and then progress to causal loops.

Public and private transport in rural areas – a mini case study

The viability of public transport in rural areas has become increasingly problematic. As more people living in villages can afford to buy and run cars, the number using public transport has declined. However, a core of users still rely on public transport to get them into the nearest town for shopping, medical services, leisure, socialising and so on. With declining passengers, timetables have been pruned to cut costs but some costs have to be passed on to passengers. Cost increases and a deteriorating service deter passengers. The increase in car usage causes congestion and safety problems not only in the villages and narrow country roads but also in the nearby towns. Figure 9.2 shows the pattern of influences relating to bus services and cars. Train and boat services have been ignored although in many rural areas such as in Scotland they would have to be included.

Exercises

3. Figure 9.2 can be converted directly to a causal diagram and thus show the dynamics of system interactions. For each of the arrowheads A to R, indicate whether it is plus (+) or minus (–).

4. Identify control loops (–) and destabilising (+) loops.

From an initial influence diagram converted to a causal model it is possible to section off parts of the system for more detailed study. For example, in the developed version of Fig. 9.2, you could separate those components that are more under the control of the bus companies and expand the causal diagram. i.e. increase the level of detail.

Comesta snacks – a mini case study

Comesta, a leading manufacturer of snack foods, has not been performing well in recent years. A high return on capital employed is

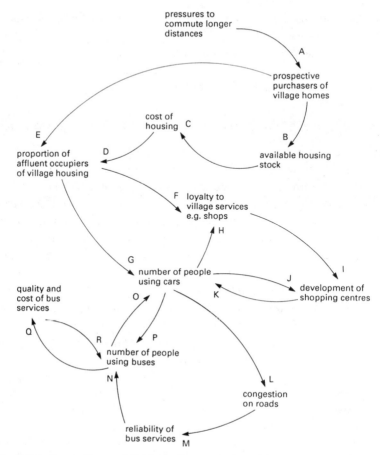

Fig. 9.2 Influences on public and private transport in rural areas.

a major strategic objective laid down by the Board of Directors. From a high return on capital of 21% four years ago, performance has fallen progressively to 10%. Market research has shown that consumers are becoming bored with the flavour of Comesta's Cruncho crisps and that they would respond favourably to salmon-and-cucumber flavour. Unlike the old flavour, the new flavour X-530 can be added to an oil base and spray-dried onto the crisps, a process that is more economical than applying the flavour on a salt base. Soon after introduction, the new process achieves an 8% reduction in production costs. Within a couple of months, however, the sales department complains that supplies of the new flavoured Crunchos from production have been as much as 15% down. It is discovered that X-530 is in short supply. The purchasing section in the production department have cancelled a long-standing arrangement with a local supplier because they felt that the supplier of X-530 knew more about spray-drying. However, the new supplier's technical expertise has not been matched by his ability to honour delivery promises. Comesta receives a 25% discount on bulk supplies of flavour from the new supplier, exactly the same as from the previous supplier. However, as a result of the cancelled bulk orders, their previous supplier will not now reinstate their discount.

Exercises

5. Draw a diagram of the pattern of influences that affect the return on capital employed in the Comesta case.

Air traffic control – a mini case study

For those of you whose holiday flights have been delayed for many hours owing to 'air traffic control problems', the civilian air traffic system may seem to be a considerable failure. Indeed, the author was so fed up with being held up for 24 hours at Luton Airport that he whiled away the time thinking about where the system was going wrong. His thoughts are presented in Chapter 13. In this Chapter, we present a mere snippet of the complex system involved.

It has been estimated by the International Air Transport Association (IATA) that the number of civilian aircraft movements over Europe during the first six months of 1988 increased by between 10 and 14% compared with the same period in 1987. These increases are part of a trend that can be traced back for several years. For example, between 1986 and 1989 the number of civilian aircraft movements involving Europe's 42 busiest airports will have increased from over 4 million to nearly 6 million. A similar pattern is discernible in the United States. However, the problems involving an increase in air traffic are not uniform and some holiday routes at weekends experience passenger increases of 20 to 30%.

Existing computer facilities in each of the European air traffic control centres are not integrated and many are outdated. Although existing computers can pinpoint the present location of an aircraft, air traffic control staff still have to anticipate the aircraft's future positions. Maintaining safety is a vital consideration in air traffic control (see Fig. 5.6 in Chapter 5, for example). Better computers and software are being planned, but in the short term aviation authorities continue to seek any available capacity gains. The air traffic system is a victim of its own success. The more successful it is in increasing capacity, the more passengers are likely to use it. Competition among airlines and tour operators contributes to the problem of too many passengers for the existing capacity. Current palliatives include: employing more controllers, increasing controller productivity, allowing more night flights, transferring military air space for use by civil aircraft, and stopping small aircraft from using major airports.

Exercises

6. Draw firstly an influence diagram of the air traffic control problem and then convert it to a causal diagram. Remember to start with a small number of components at low resolution (i.e. general topics rather than fine detail).

7. From the problem description above, what would lead you to expect counter-intuitive outcomes from current efforts to solve the problem?

Most systems, particularly human activity systems, require relative stability in order to function effectively. The presence of destabilising loops in your causal diagrams should alert you to what may be important aspects of the situation. For example, instability in one area may greatly affect the presenting problem which you are trying to solve. The precursor problem may have to be addressed first in consultation with the client set.

To round off the essential groundwork and development of understanding, the system needs to be formally defined. The system should be given an accurate name ('a system to ...') and its owner should be identified. Inputs, outputs and essential sub-systems need to be listed and a convenient reference for this is the formal system model (FSM) as described in Chapter 7. Depending on the degree of urgency, you should have carried out at least one and maybe several iterative developments within step 2. At this stage, it is likely that you as problem solver understand more about the system and its functioning than do any of the client set.

9.5 Step 3: Objectives and constraints

It would be tempting at this stage to start focussing on practical solutions. The client set will probably be anxious for results and their perception of the task may be quite unrealistic. Once trust and confidence have been established, it is common for the client set to regard the analyst as 'the expert' – someone with a magic wand who can produce a perfect solution out of a hat and for whom no practical constraint is insurmountable.

In order to avoid disappointment for the client, it is essential to draw up clearly defined objectives for the project. In essence, agreed objectives represent another contract between the client set and the analyst. Such an agreement is another test of the client set's shared world view and the likelihood of the project's success. Defining what success means to the client set, what resource limits they have and what things they cannot entertain must be done before searching for solutions.

As noted in Chapter 4, a goal is the overall target to be reached and may be expressed in more detail as a set of objectives or *measurable* results to be aimed for. Objectives and constraints will be a mixture of quantitative and qualitative things. Objectives themselves need to be systematically organised rather than simply being jotted down as a random list. They also need to be phrased unambiguously and with some clear, if broad, action in mind. In other words, each objective should have a verb in it such as 'plan', 'develop', 'implement' and 'attain'. An ideal way to make objectives manageable is to draw up an objectives hierarchy. This is an organisation chart of objectives and assumes that some objectives are subordinate to others and that some must be reached before, logically, others can be. The following example demonstrates the principle.

Information Technology in Further Education – a mini case study

In recent years, Colleges of Further Education in the United Kingdom have been forced increasingly to address pressures for change in their role, outlook and methods. For example, there has been growing concern among politicians, employers and educationalists about college attitudes towards local market conditions. The Education Reform Act of 1988 has reinforced the trend towards FE colleges adopting a more business-like approach to their activities.

You are the head of the Information Technology (IT) section in the Business Studies Department of the local FE college. Traditionally, your section has taught computing on day-release, evening classes and full-time courses. Education reform, however, is rapidly falsifying many of the traditional assumptions. Pressures are upon you to run more self-financing courses. You are also aware of competition from commercial training companies whom local employers may prefer and this threatens your day-release market. At present, computing courses are over-subscribed and so you are not unduly worried about the short-term. The danger you foresee to the viability and survival of your section lies in the longer term (5–10 years hence) when competition could be stiff and your resources not assured.

The Head of Department and other key figures share your concern because unless the IT section is viable the future of the Department is in doubt. You have been allocated £70 000 to set up a custom-built IT suite which is intended to provide better facilities for students and act as an IT training resource for the College's own academic staff.

Acting as problem owner, client and analyst rolled into one, you have started a hard systems study of the problem. You have entitled the system of interest 'the Department of Business Studies system for IT education and training'. The long-term objective is to ensure viability and survival of the IT section, but what other objectives when met contribute to it? Three parallel branches are identifiable as in Fig. 9.3. In contrast to some objective trees, in this case all three branches would have to be pursued.

Exercises

8. An obvious constraint in the above study is the amount of money allocated to providing the IT suite. What other constraints could there be?

9.6 Step 4: Strategies

In some projects, there may be only one realistic pathway to the primary objective. For example, if the objective is to reduce the number of grades of hourly paid workers in a factory from twelve to five, there is no real choice in the method used. A choice only presents

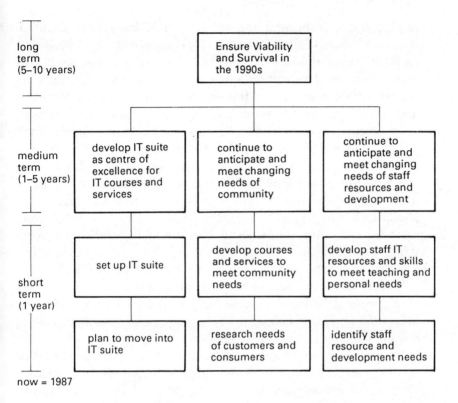

Fig. 9.3 Objectives hierarchy for the viability and survival of the IT section of the Business Studies Department in a local FE College.

itself in the technique or techniques to adopt, e.g. merging grades. In many projects, however, distinctly different options are available. For example, if the project objective is to increase company profits this could be achieved either by cutting costs or by increasing sales while maintaining prices. Both are potential options and although one may be preferred at the outset by the client set a hard systems examination of these routes may cause them to alter their view.

In complex cases, the analyst has to take the initiative in a creative search for solutions. Creative searching comprises two phases, expansion or divergent thinking followed by contraction or convergent thinking. Expansion seeks to provide as many ideas and potential strategies as time and other resources permit. Contraction seeks to classify and refine them into practical propositions. Brainstorming is a particularly good method of creative search. A detailed coverage of brainstorming is beyond the scope of this book and Geoffrey Rollinson's book entitled *Creative Thinking and Brainstorming* is a useful reference. Brainstorming seeks to encourage among a group of people (some of the client set, say) creative, almost anarchic thinking but set within a systematic framework. Even apparently ludicrous ideas should not be discouraged. For example, one group faced with the problem of how to gain access to windows on the outside of tall public buildings for cleaning and maintenance circled around and rejected a whole stock of 'standard' solutions that would have been acceptable on lower buildings. Someone then suggested training a team of flying squirrels to do the job and someone else said 'What about sending in the SAS?'. From this somewhat amusing divergence, came a much more fundamental question 'Why should any living thing, man or beast, be expected to do it?' The group proceeded to focus on automatic mechanical means for doing such work.

Sometimes, of course, you may be doing hard systems analysis on your own and brainstorming would not be appropriate. Bouncing ideas off friends and colleagues or just quiet contemplation may be helpful. Creative search is obviously an iterative process.

In devising strategies for reaching objectives, care needs to be taken in distinguishing between project objectives and client objectives. Obviously the two are connected but the client's objectives are likely to encompass the project objectives and, in effect, act as a constraint on them. For example, British Steel installed two gas-fired direct reduction plants at Hunterstone at the end of a project whose objective was to find the most efficient method of coping with extra anticipated demand for steel, given that some of the old coke-fired furnaces were at the end of their useful life. The project objective was, however, subordinate to a higher objective of the client (British Steel), namely to stay at the forefront of technology and develop some independence from coal. A further higher level client objective (possibly influenced by the government) was to improve job prospects in an area of high unemployment. Both higher objectives constrained the project objectives in terms of how they could be achieved, i.e. strategic options. The most effective method of coping with extra demand would have been to keep one of the older furnaces going until the peak in demand had passed and not to indulge in new plant. Also from the point of view of the project's objectives Hunterstone was arguably not the best location for new plant.

9.7 Step 5: Measures to assess achievement

Assessment measures are those that can be used to determine how well a strategy performs in meeting objectives. Such measures have to include quantitative ones because hard systems analysis depends on them. Typical quantitative measures are cost; return on capital employed; savings in time, materials, energy and labour; reduction in queue lengths; number of energy units produced per unit cost; miles per gallon, etc. Refer back to Chapter 3 for a discussion of quantification.

Quantifiable measures need to be framed so that a clear target level or rate is specified and a time factor built in. For example, an objective such as 'achieve a high return on capital' needs to specified as so much per cent (net or gross?) over so many years (yearly, on average, or in total?).

Qualitative measures stemming from the client set's world-view include such factors as declared policies, senior management's current interpretation of those policies, current enthusiasm for a particular technology, and their criteria for what 'acceptable' means to the organisation. Other qualitative measures relate to possible effects of implementing a solution such as effects on workforce morale and attitudes if workplace or job changes are possible, or effects on public opinion if a major building or transport programme is involved. Appropriate measures might include pressure group reaction, opportunities for staff development, corporate image projection, and so on.

Measures of assessment must be specified before proceeding to modelling and evaluation. Quantitative measures will feature prominently in these next two steps whereas qualitative measures will come into play in step 8, making a choice. As with other steps, specifying assessment measures should be done in consultation with the client set.

9.8 Step 6: Modelling

A model is a representation of something in the real world that shows either what it looks like or how it works. In hard systems studies for decision support, you will be concerned largely with numerical models. In systems engineering, you might also be concerned with physical models of engineered systems.

Typically, mathematical models are selected or devised to simulate the relationship between independent (causal) variables and the dependent variable, i.e. the measure of performance. For example, if cost is the assessment measure specified in step 5, it depends on a number of other variables and in particular fixed costs and variable costs. So at a simple level the relationship may be expressed as:

$$\text{total cost} = \text{fixed costs} + \text{variable costs}$$

or expressed with symbols
$$TC = FC + VC$$

Now fixed costs comprise things like heating, lighting, and rates whereas variable costs include the wage bill, advertising and so on. It is thus possible to refine the simple model above to include all the ingredients that contribute to TC. The result is your numerical model that simulates total cost, the assessment measure you chose in step 5 to test in step 7 each of the options on your list from step 4. Another very simple numerical model is $C = A - L$ where C is capital, A is assets and L is liabilities.

Exercises

9. Return on investment is a popular assessment measure. What is the numerical model or formula that takes into account a principal sum for investment (P), annual decimal interest rate (i), and the number of investment years (y) and allows you to predict the performance (R) of your investment in years to come?

Although some models, especially financial models, are available 'off-the-shelf', others have to be devised and refined from such sources as causal loop diagrams from step 2. A well-established financial model for comparing the relative worth of several potential investment projects is the net present value (NPV). The basis of the NPV approach is to compare the return you would get from investment in each project (e.g. launching a new product, buying in someone else's) with the return you would get by simply leaving the investment money in a bank.

For example, if you deposit £100 000 in a bank at a fixed interest rate of 10% per annum, then in five years' time that investment will be worth £161 051. Investing £100 000 in each of your possible strategies would also generate revenue in the future, for example from sales. However, will any of your options perform significantly better than simply leaving the money in the bank?

NPV is a single value or index calculated from adding up the present values (PVs) for each year of the investment period. The PV represents the value at today's prices. Thus, from the figures above, £161 051 in five years' time would only be worth £100 000, i.e. its present value. If the project cost was only £80 000, and achieved a return in five years' time of £161 051 then the return on investment would be better than the bank's.

The PV for each year is calculated from the formula $PV = I/(1 + i)^n$ where I is the investment sum, i is the annual interest rate (as a decimal), and n is the period of investment. However, project cash has a flow to it – either income is greater than expenditure (positive cash flow) or vice-versa with a negative cash flow. The net result of adding total income (+ values) to total expenditure (– values) for each year is the net cash flow, and it is this figure that is used as I in the PV formula. Table 9.1 shows how PV is calculated for each year of investment to give a discounted cash flow.

Table 9.1 Comesta investment project: Cruncho crisps – new flavour X-530

Item	Cash flow (£ '000) in year					
	0	1	2	3	4	5
1. Net income			45	250	432	335
2. Net expenditure	–133	–101	–380	–231	–100	–95
3. Net cash flow (1 + 2)	–133	–101	–335	+19	+332	+240
4. PV @ 10% (3 as I)	–133	–92	–227	+14	+227	+149

5. NPV @ 10% (sum of all PVs) = –62 (£ '000)
6. Total income less total expenditure = 1062 – 1040 = +22 (£ '000)

Note how although there is an apparent net profit of £22,000 the NPV is negative.

The Comesta example has been greatly simplified and in practice you would itemise the main contributions to net income and net expenditure. You would also include the effects of payments or rebates of corporation tax. Financial modelling techniques can be quite powerful; even using elementary computer software such as spreadsheets enable 'what if?' projections to be made. Knowledge-based computer programs also enable aspects of a problem to be modelled in conceptual terms as an aid to definition.

The negative sign to an NPV is not good news for it indicates that the investment will make a loss. However, five years is a relatively short period over which to judge performance and extension of the cash flow table might well produce a healthy positive NPV.

It is worth noting that in calculating PVs companies normally do not use the current minimum lending rate or business loan rates. Instead, they set a 'hurdle rate' which is several percentage points above business interest rates. The higher gearing is to take account of the desired profit and is indicative of management's world-view.

9.9 Step 7: Evaluation

Any model to be used for evaluating potential options must have been 'dry run' tested in step 6 to ensure that it is sufficiently accurate for its purpose. Once confidence in the model is established, different sets of likely values for the independent variables can be used on a 'what if?' basis to compute corresponding values of the performance measure. The result might be a set of NPV figures as outlined in the previous section or some other set of data that enables the value of each strategy to be compared.

You and the client set need to be aware that such numerical evaluations do not represent certainty; they are rules-of-thumb indicators that, say, one or two routes look more promising than others. The degree of uncertainty can be controlled by techniques such as decision analysis which introduces client set estimates of the probabilities that, say, certain market factors will prevail over the period concerned. For example, the NPV for Comesta's new flavoured crisps might be £10m at 10% growth rate and £25m at 18% growth rate, but the client set reckon that the chances of 18% growth rate are 0.7. Combining these data $(0.7 \times 25) + (0.3 \times 10)$ gives an expected monetary value (EMV) of £20.5m. Comesta could examine its other options in a similar manner so that their EMVs may be compared.

9.10 Step 8: Making a choice

As useful as quantitative modelling is, it can only be regarded as a support for decision making. The client set and client organisation will have qualitative objectives to consider and there may well be hidden agendas which you, as analyst, know little or nothing about. Quantitative evaluation suggests an optimum solution but qualitative evaluation by the client may result in a lower performance route being selected. This is a 'satisficing' selection, i.e. what is considered to be the most satisfactory on all counts rather than the numerical best.

Various scales and weighting factors such as those used in the Kepnor Tregoe approach can overcome some of the incompatibility of quantitative and qualitative factors at the selection stage. A formal presentation to the client set supported by a detailed project report is a normal requirement at this stage.

9.11 Step 9: Implementation

Many problem solvers feel that their work has finished at the end of step 8. However, planning for implementation and the changes involved should have been clearly in the minds of analyst and client set from the beginning of the study. This is especially important where major organisational or technological changes are likely. Where employees are going to be affected directly, it is better to keep them informed, or maybe even involved, from an early stage. This will help avoid suspicion and worry about 'secret working parties' planning the demise of this or that group in the name of cost-cutting and efficiency. Joint working parties and steering committees, for example, can aid communication and dispel unsubstantiated rumours in addition to helping to plan the implementation phase.

9.12 Summary

A hard systems approach to managerial problem solving guarantees nothing. Where there is general agreement about the nature of a quantifiable problem and the goal to be reached, it provides a rational tool for finding a range of solutions and aiding the selection of the most satisfactory one from the client's point of view. Chapter 10 provides two worked case studies using hard systems analysis.

9.13 Suggested answers to exercises

1. When solutions work counter-intuitively or produce unexpected new problems, it suggests that the problem solver did not understand the systemic context of the problem. The solution when introduced will affect other components within the system and their interaction. If these components and their relationships are inadequately known and understood, the solution is likely to be poor.

2. Hard systems analysis is unlikely to be successful if the client set does not have a shared world-view or if key individuals are 'at each other's throats'. When such mismatches and tensions are revealed, the analyst would be well advised to switch to the soft systems approach and return to the hard method once the soft issues had been explicitly confronted and dealt with. The art of switching methods is covered in Chapter 15.

3. A (+); B (–); C (–); D (+); E (+); F (–); G (+); H (–); I (–); J (+); K (+); L (+); M (–); N (+); O (–); P (–); Q (–); R (–).

4. There are five destabilising loops:
   Q-R-Q
   O-P-O
   L-M-N-O-L
   K-J-K
   K-H-I-K

5. Figure 9.4 shows the influence diagram of factors affecting Comesta's return on capital employed. Your version may be different and even superior to ours. What counts is *your* perspective. You can always refer back to the client set to test the accuracy of your diagram.

6. The causal diagram of the air traffic control problem is shown in Fig. 9.5. As you know, usually thick lines are used to denote major influences or causal links. In Fig. 9.5, however, we have used thick lines to show the simple diagram we started with before expansion.

7. Counter-intuitive outcomes are to be expected because the highly

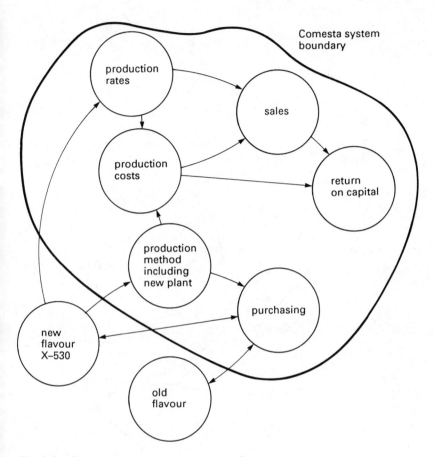

Fig. 9.4 Influences on return on capital at Comesta.

complex system and the nature of the problem is inadequately understood. Short-term solutions (palliatives) are being implemented as common-sense responses simply to keep the system operating at all.

8. Staff numbers and staff time are two obvious constraints. Teaching staff can only be in one place at a time, and if your IT lecturers are teaching in a classroom they cannot be visiting local employers to carry out market research, for example.

9. The model is $R = P \times (1 + i)^{y}$. Of course, this formula for the level of return assumes a fixed interest rate and ignores tax liability.

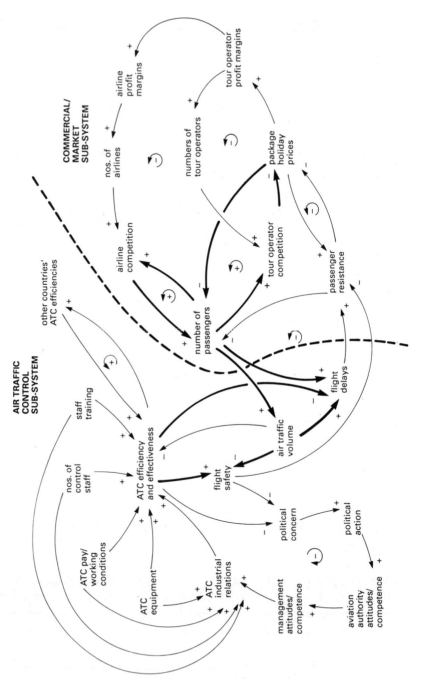

Fig. 9.5 Causal loop diagram of the air traffic control problem in relation to holiday traffic (second iteration). *Note:* thick lines denote initial diagram and *not* main causal links.

Chapter 10
Hard Systems Case Studies

10.1 Introduction

Previous chapters (3, 4 and 9) have given you a grounding in hard systems ideas and techniques. In this chapter, we provide two case studies that demonstrate rather different uses of hard systems analysis. The first case, Redwood Breweries, concerns decisions about business strategy in a changing market environment facing a small, regional brewery. We are indebted to John Lewington, for providing background material for this case study.

The second case, Datatrieve, concerns the development of a customised software package for a client's information processing needs. Whereas the Redwood case shows HSA being used for decision support, this case is about software design and to a large extent falls in the systems engineering domain of HSA. Emphases are different and the Datatrieve study shows that HSA can be used quite flexibly.

10.2 Redwood Breweries – a study of strategic decision making in a rapidly changing market environment

Redwood Breweries is a small, traditional brewing firm in the West Country. Over its 150-year history, Redwood has enjoyed respect in the region for its beers. It currently runs two breweries, one at Bodmin and the other at Yeovil, which support Redwood's 390 pubs spread throughout Devon, Cornwall, Somerset, Dorset, Avon and Wiltshire. The West Country is a magnet for summer tourists and seasonal demand is evidenced by the fact that some 90% of Redwood's profits are generated in the summer.

As a traditional 'real ale' brewer, Redwood still uses beechwood fermentation tanks for most of its production but even Redwood has had to recognise the change in drinking patterns since the early 1970s, especially among 18–24 year olds. Whereas it used to brew only 'real ale' bitters, milds and stouts, now 25% of production is lager brewed in stainless steel tanks. Anti drink-and-drive laws have seem a massive R&D initiative in the brewing industry to develop low-alcohol beers and Redwood has had to follow suit. However, its product mix is still dominated by real ale at 50% of total output.

In the last financial year, Redwood produced 110 000 barrels divided roughly equally between Bodmin and Yeovil. The Yeovil site also has a bottling plant which handles 12 000 barrels. Both plants also bottle mineral water of which 350 000 gallons in total were output last year. In addition to distribution direct from the two breweries, Redwood also has distribution depots at Barnstaple and Chippenham. While Redwood is strong in the region, its assets, capacity and output are small compared with major UK brewers.

Like most of the major brewers, Redwood has tried to develop into the hotel trade. This move was prompted by the growth in leisure spending and the opening up of the West Country by the M4 and M5 motorways. Tourists and short-break holiday makers from London and the Midlands can now be in the West Country in under two hours. Redwood set up a subsidiary company in 1980, initially to purchase two seafront hotels, the large Duchesne Hotel in Exeter and a smaller summer season hotel, The Vista, in Truro. After costly refurbishment, the hotels reopened during the recession of the early 1980s. Poor profits from the hotels have led to their recent sale.

Redwood's Board of Directors has seen several recent changes. David Redwood-Curry retired after twenty years as Chairman and nearly 50 years as a Director. The new Chairman is John Hutchinson, whose former post as Finance Director is now held by Alan Walker previously Group Accountant with Coopers, a major UK brewer. Coopers own 25% of Redwood shares. Alan Walker is a very experienced accountant with an MBA from Warwick Business School and has a reputation for applying modern management techniques.

The arrival of Alan Walker was influenced by Coopers who last year were both pleased but also a little dismayed to learn that following a revaluation of Redwood's properties a surplus of £24m over the previous book value had been revealed. In essence, Redwoods had become 'asset rich' in terms of property values but 'profit poor' in terms of return on capital employed in its revenue operations. As a major shareholder, Coopers were bound to take action to protect their return on investment. The UK property boom in general had affected the value of Redwood's property portfolio. In addition, however, Redwood's pubs tend to command premium locations in villages where residential property is in high demand from retired couples, families seeking second or holiday homes, and 'yuppies' willing to commute by motorway to jobs in Bristol, London and the M4 high-tech corridor. Thus, the value of Redwood's pubs on the property market is well in excess of their commercial value to the company, a situation that is unlikely to change in the foreseeable future.

Last year, Redwood spent £1.9m as capital expenditure on refurbishing pubs, new vehicles, and minor upgrading of brewing capacity at its Yeovil brewery. This expenditure was financed by retained earnings and the sale of four pubs. Redwood has loans of only £0.5m against assets of more than £40m, a reflection of historically low borrowings.

In addition to changing drinking habits of 18–24 year olds, Redwood's market is changing rapidly on a number of dimensions. Not only is the residential population of West Country villages changing but so too is the transient holiday population. Cheap package holidays in sunny countries have drawn away domestic holiday makers. Whereas ten years ago holiday makers decided well in advance to spend two or three weeks annual holiday in the West Country, nowadays the trend is for holidays abroad with perhaps a short second holiday in the UK. Ad hoc weekend break holidays have become popular. Although demand for small hotel and guest house accommodation has remained steady, short camping and caravan breaks are on the increase. The unpredictability of English weather has added to the uncertainties about the future of tourism in the West Country, the main plank of Redwood's markets.

In the face of market volatility, Alan Walker is keen to reduce Redwood's vulnerability and intends to embark on a rapid growth strategy. The Board has agreed to an independent study by Axis Management Consultants of how rapid growth could best be achieved

Step 1: Groundwork

The Axis analyst had an initial two-hour meeting with Alan Walker to establish rapport and to obtain essential facts and figures about Redwood. Alan handed over a number of reports and files. The analyst also probed discreetly Alan's relationship with the rest of the Board.

Did he enjoy their full confidence or was he seen as a 'new broom' to be tolerated but not allowed full rein? What did Alan see as the Chairman's attitude, bearing in mind that John Hutchinson previously held the Finance Director's position? Alan's responses indicated that there were no divisions within the Board as to the need for rapid growth and a new strategy to achieve that goal.

The analyst outlined a programme for the study over three months and agreed with Alan when interim reports would be submitted to him and when a final presentation to the Board would be made. Alan would personally authorise payment of staged invoices for the study. This convinced the analyst that Alan was both the problem owner and the client and, importantly, appeared to hold sway over the Board's ultimate decision. Nevertheless, the analyst needed to interview the remainder of the client set to assure himself that there were no hidden world-views that would work against an eventual decision for concerted action. At this stage, the analyst considered the client set to comprise Redwood's Board but he held in reserve the possibility of others with whom the study might have to be credible. For example, major shareholders, Redwood's bankers or other sources of finance might need to be 'sold' on recommendations for change in Redwood's market strategy.

The analyst interviewed each member of the Board separately. It became clear that although there were 'old guard' and 'new guard' directors there was no fundamental disagreement about the need for a new approach to the business. Although as an old family firm in a traditionally conservative industry there was a natural tendency to evolve slowly, the client set were enlightened enough to recognise the need for change. Far from antagonism towards Alan Walker, he was described in warm terms and his enthusiasm, up-to-date business skills and experience in the industry were clearly valued. He was seen as an insider.

Before starting on systems analysis, the analyst requested a brief meeting with the Board to review the nature of the study and the overall goal (rapid growth with profitability), and to confirm that Redwood was seeking compatible rather than radical non-brewing solutions. Although he never used the term 'world-view', the analyst also sought confirmation that the client set's world-view was essentially that they were the inheritors of an ancient social industry for providing alcoholic refreshment products and convivial meeting places where people could enjoy them. The client set felt an obligation to maintain tradition but also recognised an obligation to 'change with the times' and so continue to meet market needs and wants. They regarded their beers with pride but were not 'real ale' campaigners. They would strive to achieve excellence in whatever refreshment products and related leisure services the public demanded. In short, although their attention had been drawn to an imbalance between assets and profits, the client set saw this as a short-term problem whose resolution presented an opportunity to strengthen their position in the product market. While they were not risk aversive, they were used to evolutionary change and were unlikely to agree to radical high-risk proposals. Diversifying into other leisure areas such as bingo halls and theme parks would not fit their own identity or the image they wanted to maintain. Brewing and pubs were central to their past, present and future.

Step 2: Awareness and understanding

The analyst summed up his impressions of the problem to be solved as one of finding viable ways to improve Redwood's market position as a traditional, regional brewer in a rapidly changing leisure industry and market. Redwood's Board were committed to fulfil opportunities for expansion that would also solve a temporary problem of low profitability and stabilise their position vis-à-vis a volatile market. The analyst's own commitment as a management consultant was to help the client set achieve their overall objectives by presenting viable choices within an agreed three-month study period. The analyst drew a spray diagram of the situation as shown in Fig. 10.1. It should be noted that at the time the Monopolies and Mergers Commission had not announced their investigation into the brewing industry, especially the tied house system that is alleged to hide brewers' true profits. Obviously, any potential threat of government intervention would have had implications throughout the study.

Was the commitment of Redwood's Board in the person of Alan Walker a systemic one? Would a hard systems study be worthwhile, for example, or would a formal problem-solving exercise be adequate? Since Redwood was subject to a complex and volatile market situation with many variables operating, the analyst concluded that formal problem solving would be inadequate. Was the exercise seeking some achievable, measurable goal? Was there some definite action in mind at the end of the study? Was Alan Walker a key figure in the decision

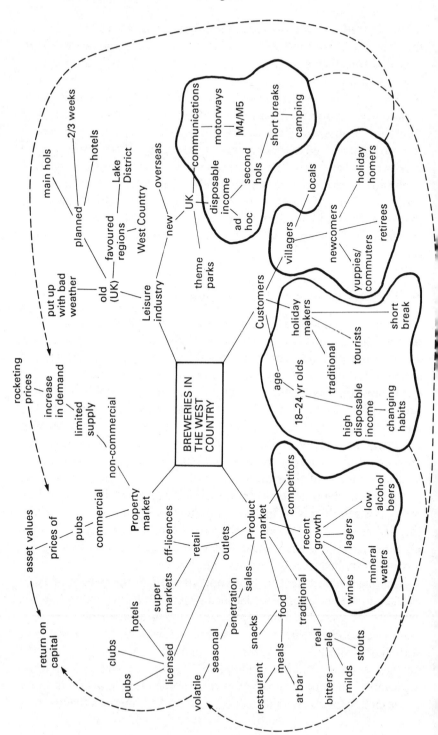

Fig. 10.1 Spray diagram of the Redwood Brewery situation (first iteration).

A fruitful area in Redwood's situation: market mix

A potential system: the Redwood system for achieving a market mix to provide rapid growth with profitability in the brewing and leisure markets.

(a) first iteration

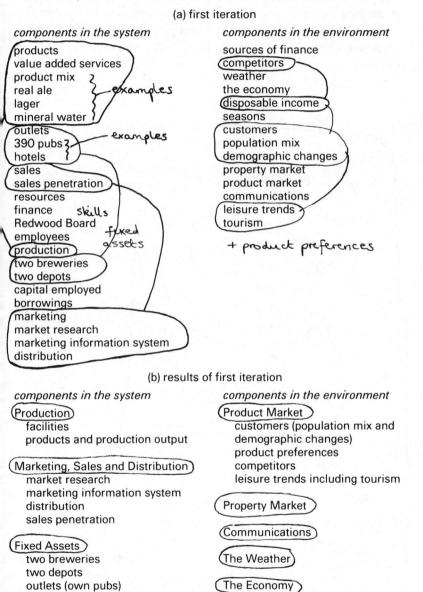

components in the system

products
value added services
product mix ⎫
real ale ⎬ examples
lager ⎪
mineral water ⎭
outlets ⎫ examples
390 pubs ⎬
hotels ⎭
sales
sales penetration
resources
finance   skills
Redwood Board   fixed
employees   assets
production
two breweries
two depots
capital employed
borrowings
marketing
market research
marketing information system
distribution

components in the environment

sources of finance
competitors
weather
the economy
disposable income
seasons
customers
population mix
demographic changes
property market
product market
communications
leisure trends
tourism

+ product preferences

(b) results of first iteration

components in the system

Production
  facilities
  products and production output

Marketing, Sales and Distribution
  market research
  marketing information system
  distribution
  sales penetration

Fixed Assets
  two breweries
  two depots
  outlets (own pubs)

Resources
  finance
  Redwood Board's skills
  employees' skills

components in the environment

Product Market
  customers (population mix and
  demographic changes)
  product preferences
  competitors
  leisure trends including tourism

Property Market

Communications

The Weather

The Economy

Sources of Finance

Fig. 10.2 Separation of a system relevant to Redwood's problem.

process? On all counts, it was clear that the situation was systemic and that hard systems analysis was warranted.

Using the spray diagram as reference, the analyst then separated out what seemed to be a key system, namely Redwood's system for achieving a market mix to provide rapid growth with profitability in the brewing and leisure markets. Figure 10.2 shows one iteration of component separation and an adjustment of resolution. From this separation, it was possible to draw a system map as in Fig. 10.3.

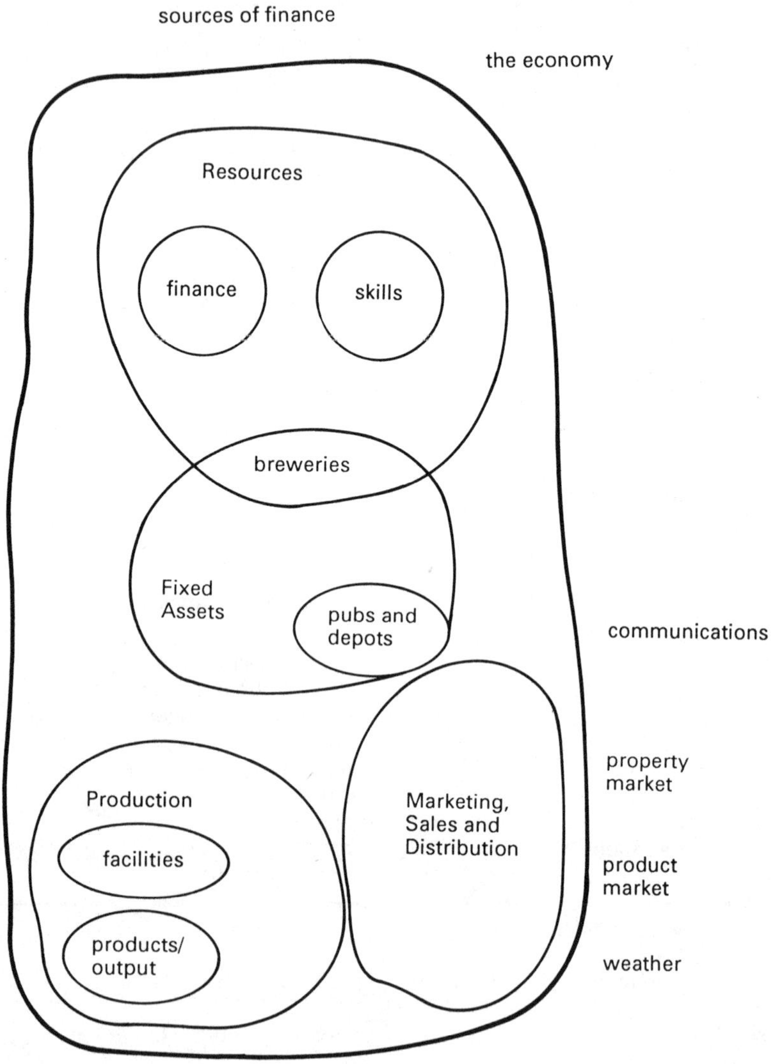

Fig. 10.3 System map of the Redwood system.

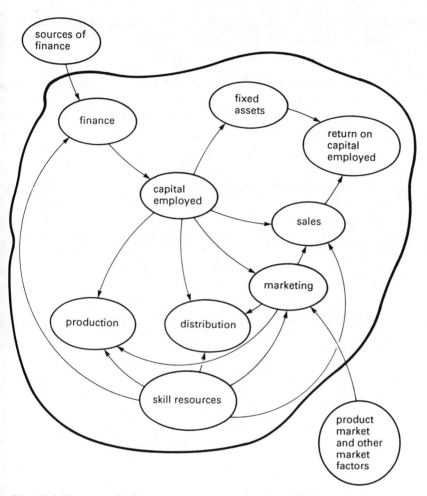

Fig. 10.4 Diagram of influences on return on capital at Redwood.

The client set's overall objective of growth with profitability, i.e. an acceptable return on capital employed is influenced by a number of factors under Redwood's control:

- finance
- fixed assets (especially property)
- capital employed (on revenue operations)
- skill resources (especially at Board level)
- production (production facilities, capacities etc.)
- sales (volumes, values, penetration etc.)

- marketing (market research, market information system, product positioning, product portfolio etc.)
- distribution (methods, capacities etc.)

Fig. 10.4 shows how these influences affect return on capital employed.

The Axis consultant began to view the growth problem as one of how to ensure that Redwood used its capital investment to increase its share(s) of the market(s) available to it. If its market shares were small and sales were sluggish, return on capital employed was bound to be poor. Figure 10.5 shows his initial attempt to address the main dynamic variables; Fig. 10.6 shows an expanded version. These diagrams have no control loops on the supply side of the market and so if Redwood failed to increase its own sales and market shares then competitors would almost certainly make their position stronger at its expense.

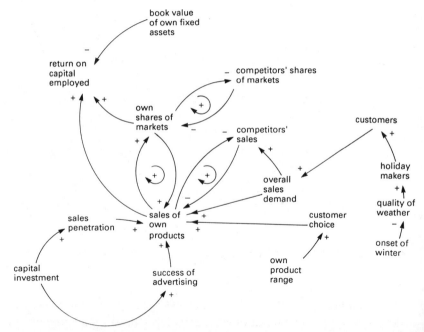

Fig. 10.5 Causal loop diagram of variables affecting return on capital employed at Redwood (first pass).

Step 3: Objectives and constraints

The twin threats of increasing competition and inconsistent but overall static demand in the West Country convinced the consultant that Redwood had to increase its overall sales, whether at the expense of

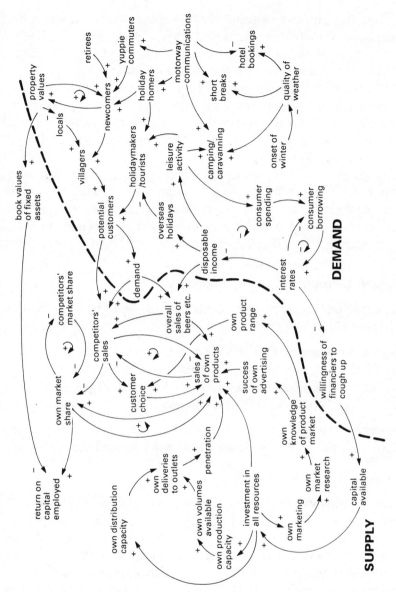

Fig. 10.6 Causal loop diagram of variables affecting capital employed at Redwood (third iteration).

existing competitors or via the creation of new markets or whatever. The overall objective of growth with profitability was confirmed. Sub-objectives were to reduce Redwood's reliance on its traditional market and to achieve a healthier balance between profits and assets.

Time was an obvious constraint. Although Redwood was not at crisis point, the situation demanded fairly urgent action. The consultant's study alone would take three months and implementing a decision could well take a further six to nine months.

The Board had already signalled their reluctance to go for a non-brewing solution and so this limited the possible ways to reach the objectives. A further possible constraint was the availability of capital to finance growth. Although Redwood had sold assets when it thought it necessary, it was not going to sell off large parts of its pubs and other premises as these were an integral and necessary part of its sales operation. However, if it had to borrow money, a rapid rise in interest rates could adversely affect cash flow and servicing of the loan. Redwood would therefore be considering very carefully how the growth strategy would be financed.

Step 4: Strategies to meet objectives

The consultant identified four possible growth strategies on the basis of a 10-year plan:

(A) Go for a larger share of their current West Country market.
(B) Create new product market(s) allied to existing ones.
(C) Enter/take similar markets outside West Country.
(D) Move into different product markets.

Strategy D was eliminated as it involved diversification by acquiring a non-brewing company, something that Redwood did not want to do.

Strategy A: acquire Dartstone Taverns

Strategy A involved the acquisition of a smaller brewer in Redwood's market area as a means of increasing Redwood's market penetration and therefore sales. The purchase of Dartstone Taverns and its 50 pubs would cost £9.0m. The consultant estimated that this acquisition would generate a cash-flow of £2.0m per annum growing at 15.0% per annum on the assumption that it was an even chance that the West Country holiday market would remain strong. A worst case estimate if the holiday market was poor would be an annual cash-flow of £0.6m. In the latter case, a decision would have to be made after two years whether to retain or sell the Dartstone brewery. If sold, the Dartstone site would realise £1.5m and would produce operational savings of £0.4m per annum.

Strategy B: develop fast food outlets in main Redwood pubs

This strategy involved developing 50 fast food outlets at an estimated investment cost of £4.0m, including refurbishment. If successful (0.65 probability), this strategy would produce a cash flow of £1.0m per year growing at 15% per annum. Success would lead to a decision in three years' time either to invest in 50 more outlets at a cost of £3m or to retain the first 50 and invest in a different area of the business. If unsuccessful, the fast food chain would generate an estimated £0.3m per annum. As an alternative, the chain could be sold after three years for £1.5m or, if market conditions were poor, for £1.0m (0.5 chance of either selling option).

Strategy C: acquire Midland Breweries

Although this would be the simplest way to enter new markets, the takeover could either go smoothly or there could be a battle with competitive bids pushing up the price. The estimated cost of a smooth acquisition is £10m whereas with a competitive bid the cost could rise to £15m. The chances of either are both 0.4. If the bid failed, there would still be a capital gain of £2m on Redwood shares.

In the event of a successful takeover of Midland, Redwood would have to decide about the future of its Bodmin brewery which would probably be uneconomic in the enlarged company. The Bodmin site would fetch £2.5m at the end of the first year. Closure would also yield a saving of £0.75m per annum. In any event, demand for Midland beers could be either high or low and the effects of both eventualities (at 0.5 probability) would have to be calculated.

Step 5: Assessment measures

The Axis consultant considered what the most appropriate measures would be to estimate how well each of the three proposed strategies would perform in relation to the defined objectives. The overall objective was growth with profitability. Therefore, quantitative measures would be needed to predict return on investment. The most appropriate measures would be present value (PV) and net present value (NPV) as outlined in Chapter 9. Once these values had been calculated, expected values (EVs) could be computed to provide financial indices for each strategy.

Quantitative measures are essential but alone they provide no indication of intangibles relevant to the all-important world-view of Redwood's Board. Maintaining Redwood's identity as a traditional West Country brewer and its assumed image in the West Country as a solid, dependable employer would be important. Each strategy would have to be tested against these and other qualitative measures. For example, patterns in consumer taste, both regional and over time, needed to be relatively stable. Qualitative measures from market research therefore would have to be taken into account.

Step 6: Modelling

Quantitative modelling involved the creation of discounted cash flows for each option within each of the three strategies. Cash flows were discounted at 15% at Alan Walker's insistence. Although commercial lending rates were at 8%, he was seeking a clear net return of 5% and so a 13% overall return would be needed. Lending rates had been falling slowly but there was always the possibility that they would rise again. Allowing for a 2% rise in lending rates, he had specified a hurdle rate of 15% return.

The following discounted cash flows summarise how present values and NPVs were calculated. PVs for each year (right-hand column) were calculated by multiplying the cash flow value for that year (second column from the left) by a factor for 15% discount (third column from left) obtained from published tables. The PV of the 10-year investment was calculated by adding up the PVs in the right-hand column. Subtracting the investment gives the NPV.

Strategy A: acquire Dartstone Taverns

Strong holiday market and retain Dartstone brewery

Year	Cash flow (£m)	Discount factor	Present value (£m)
1	2.0000	0.8696	1.7392
2	2.1000	0.7561	1.5878
3	2.2050	0.6575	1.4498
4	2.3153	0.5718	1.3239
5	2.4310	0.4972	1.2087
6	2.5526	0.4323	1.1035
7	2.6802	0.3759	1.0075
8	2.8142	0.3269	0.9200
9	2.9549	0.2843	0.8401
10	3.1027	0.2472	0.7670

Present value of 10-year investment:	11.9475
Less investment:	(9.0000)
Net present value (£m):	2.9475

Poor holiday market and retain Dartstone brewery

Year	Cash flow (£m)	Discount factor	Present value (£m)
1	0.6000	0.8696	0.5218
2	0.6000	0.7561	0.4537
3	0.6000	0.6575	0.3945
4	0.6000	0.5718	0.3431
5	0.6000	0.4972	0.2983
6	0.6000	0.4323	0.2594
7	0.6000	0.3759	0.2255
8	0.6000	0.3269	0.1961
9	0.6000	0.2843	0.1706
10	0.6000	0.2472	0.1483

Present value of 10-year investment:	3.0113
Less investment:	(9.0000)
Net present value (£m):	–5.9887

Poor holiday market and sell Dartstone brewery after two years

Year	*Cash flow (£m)	Discount factor	Present value (£m)
1	0.0000	0.8696	0.0000
2	0.0000	0.7561	0.0000
3	0.4000	0.6575	0.2630
4	0.4000	0.5718	0.2287
5	0.4000	0.4972	0.1989
6	0.4000	0.4323	0.1729
7	0.4000	0.3759	0.1504
8	0.4000	0.3269	0.1308
9	0.4000	0.2843	0.1137
10	0.4000	0.2472	0.0989

Present value of 10-year investment:	1.3573
Less investment:	0.0000
Net present value (£m):	1.3573

* Residual values have been ignored for simplicity.

The sale of the Dartstone brewery after two years at £1.5m would be worth £1.5m × 0.7561 or £1.1341m (discounted at 15%). Therefore the present value of investment under poor market conditions and the sale of Dartstone brewery would be:

sale of brewery site (PV):	1.1341
operational savings (PV):	1.3573
poor market cash flow (PV):	3.0013
present value of 10-year investment:	5.4927

Taking the original investment of £9m into account, gives an NPV of −3.5073 £m. Thus, in a poor market, selling the Dartstone brewery would be better than keeping it even though it would merely be reducing the loss. The negative sign of the NPVs indicate that in poor market conditions the Dartstone option would be risky, but even in a strong market the NPV is only weakly positive.

Strategy B: fast food outlets

If successful and 50 outlets retained

Year	Cash flow (£m)	Discount factor	Present value (£m)
1	1.0000	0.8696	0.8696
2	1.1000	0.7561	0.8317
3	1.2100	0.6575	0.7956
4	1.3310	0.5718	0.7611
5	1.4641	0.4972	0.7280
6	1.6105	0.4323	0.6962
7	1.7716	0.3759	0.6659
8	1.9487	0.3269	0.6370
9	2.1436	0.2843	0.6094
10	2.3579	0.2472	0.5829

Present value of 10-year investment:	7.1774
Less investment:	(4.0000)
Net present value (£m):	3.1774

If successful and extra 50 outlets installed

Year	Cash flow (£m)	Discount factor	Present value (£m)
1	1.0000	0.8696	0.8696
2	1.1000	0.7561	0.8317
3	(1.7900)*	0.6575	(1.1769)
4	2.5410	0.5718	1.4529
5	2.7951	0.4972	1.3897
6	3.0746	0.4323	1.3291
7	3.3821	0.3759	1.2713
8	3.7203	0.3269	1.2162
9	4.0923	0.2843	1.1634
10	4.5015	0.2472	1.1128

Present value of 10-year investment:	9.4598
Less investment:	(4.000)
Net present value (£m):	5.4598

*Income of £1.21m less purchase of 50 more outlets at £3m gives a cash flow of – £1.79m.

Thus, the larger NPV in the latter case indicates that if the first 50 fast food outlets are successful it would be better to install an extra 50 than simply to retain the initial ones.

If the market is sluggish but the initial outlets are retained

Year	Cash flow (£m)	Discount factor	Present value (£m)
1	0.3000	0.8696	0.2609
2	0.3000	0.7561	0.2268
3	0.3000	0.6575	0.1973
4	0.3000	0.5718	0.1715
5	0.3000	0.4972	0.1492
6	0.3000	0.4323	0.1297
7	0.3000	0.3759	0.1128
8	0.3000	0.3269	0.0981
9	0.3000	0.2843	0.0853
10	0.3000	0.2472	0.0742

Present value of 10-year investment:	1.5058
Less investment:	(4.0000)
Net present value (£m):	−2.4942

If the outlets are not successful and they are sold after three years, they will have produced income of $(0.2609 + 0.2268 + 0.1973) = 0.6850$ £m. If market conditions are good, the sale price is estimated at £1.5m which has a present value of $(1.5 \times 0.6575) = 0.9863$ £m. Under bad market conditions, the estimated sale price of £1m would have a present value of 0.6575 £m.

Sale of outlets under good market conditions:

income over 3 years:	0.6850
selling price:	0.9863
present value (£m):	1.6713

Sale of outlets under bad market conditions:

income over 3 years:	0.6850
selling price:	0.6575
present value (£m):	1.3425

Strategy C: acquire Midland Breweries

If high demand for Midland beers and Bodmin brewery retained

Year	Cash flow (£m)	Discount factor	Present value (£m)
1	1.3000	0.8696	1.1305
2	1.5000	0.7561	1.1342
3	1.7000	0.6575	1.1178
4	1.8000	0.5718	1.0292
5	2.0000	0.4972	0.9944
6	2.1000	0.4323	0.9078
7	2.1000	0.3759	0.7894
8	2.1000	0.3269	0.6865
9	2.1000	0.2843	0.5970
10	2.1000	0.2472	0.5191

Present value of 10-year investment: 8.9059
Less investment: (10.0000)
Net present value (£m): −1.0941

If low demand for Midland beers

Year	Cash flow (£m)	Discount factor	Present value (£m)
1	0.5000	0.8696	0.4348
2	1.0000	0.7561	0.7561
3	1.1000	0.6575	0.7233
4	1.1000	0.5718	0.6290
5	1.3000	0.4972	0.6464
6	1.4000	0.4323	0.6052
7	1.5000	0.3759	0.5639
8	1.5000	0.3269	0.4904
9	1.5000	0.2843	0.4265
10	1.5000	0.2472	0.3708

Present value of 10-year investment: 5.6464
Less investment: (10.0000)
Net present value (£m): −4.3536

If Bodmin brewery sold

Year	Cash flow (£m)	Discount factor	Present value (£m)
1	0.0000	0.8696	0.0000
2	0.7500	0.7561	0.5671
3	0.7500	0.6575	0.4931
4	0.7500	0.5718	0.4289
5	0.7500	0.4972	0.3729
6	0.7500	0.4323	0.3242
7	0.7500	0.3759	0.2819
8	0.7500	0.3269	0.2452
9	0.7500	0.2843	0.2132
10	0.7500	0.2472	0.1854

Present value of 10-year investment:		3.1119
Less investment:		0.0000
Net present value (£m):		3.1119

If the Bodmin brewery were sold at the end of year 1 for an estimated £2.5m, the present value of the sale would be $(2.5 \times 0.8696) = 2.1740$ £m. This value together with operational savings of 3.1119 £m gives a PV of 5.2859 £m.

(a) high demand for Midland beers/retain Bodmin brewery:
    present value: 8.9059 £m

(b) high demand for Midland beers/sell Bodmin brewery:
    present value: (8.9059 + 5.2859)   14.1918 £m

(c) low demand for Midland beers/retain Bodmin brewery:
    present value: 5.6464 £m

(d) low demand for Midland beers/sell Bodmin brewery:
    present value: (5.6464 + 5.2859)   10.9323 £m

Step 7: Evaluation

Evaluation seeks to draw meaning from the data emanating from the modelling stage. What the data mean has to relate to the client set: what do Alan Walker and his colleagues on Redwood's Board make of all the PV data for the three strategies and various options within them *taking uncertainty into account?*

Estimates had already been made for the likelihood of certain outcomes occurring if a particular strategy were to be followed. In order to estimate the effects of such uncertainties on present values, the Axis consultant first constructed a decision analysis tree to clarify the relationship between the strategic options as in Fig. 10.7.

In Fig. 10.7, reading from the root on the left each square represents the client's decision or choice of action in the options that follow on the right of the square. Each circle represents a point of uncertainty in the possibilities that follow to the right of the circle. For example, following the Dartstone option the chances of a strong market are judged to be no better than 50/50. If the market is strong, Redwoods would almost certainly keep Dartstone's brewery (i.e. probability of 1.0). If the market is weak, Redwoods could either retain or sell Dartstone's brewery. Each path would produce different financial outcomes.

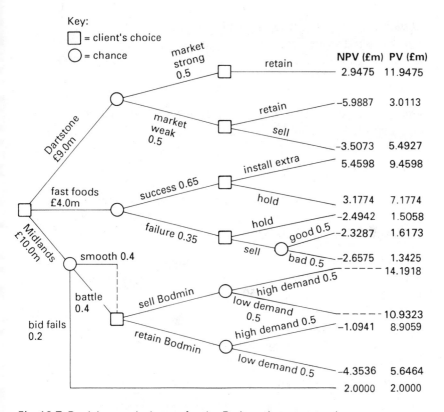

Fig. 10.7 Decision analysis tree for the Redwood strategy options.

The next step was to calculate the expected values (EVs) for each likely pathway or combination as follows.

Strategy A: acquire Dartstone Taverns

The estimated chance of the holiday market remaining strong is 50%. Thus, the expected value of the Dartstone investment under strong market conditions would be 0.5 × the cash flow PV:

(a) expected value (strong market):
   0.5 × 11.9475 = 5.9738 (£m)

(b) expected value (weak market/retain Dartstone brewery):
   0.5 × 3.0113 = 1.5057 (£m)

(c) expected value (weak market/sell Dartstone brewery):
   0.5 × 5.4927 = 2.7464 (£m)

Option 1: acquire Dartstone and retain its brewery
expected value (a + b):	7.4795
less investment cost:	(9.0000)
expected value (£m):	−1.5205

Option 2: acquire Dartstone and sell its brewery if necessary
expected value (a + c):	8.7202
less investment cost:	(9.0000)
expected value (£m):	−0.2798

Thus, the Dartstone acquisition appears to be a loss-making venture.

Strategy B: fast food outlets

If failure occurs, the expected value on selling is 0.5 × (1.6713 + 1.3425) = 1.5069 £m. This suggests that selling would be only marginally better than retaining the outlets (1.5057 £m).

Option 1: if successful install more outlets/if unsuccessful then sell
expected value: (9.4598 × 0.65) + (1.5069 × 0.35) =	6.6764
less investment cost:	4.0000
expected value of option:	2.6764

Option 2: if successful hold position/if unsuccessful then sell
expected value: $(7.1774 \times 0.65) + (1.5069 \times 0.35) = 5.1927$
less investment cost: 4.0000
expected value of option: 1.1927

Strategy C: acquire Midland Breweries

Option 1: sell Bodmin brewery
expected value: $0.5 (14.1918 + 10.9323) = 12.5620$
operational savings + selling price: 5.2859
expected all-up value: 17.8479 £m

expected value (smooth takeover):
$0.4 (17.8479 - 10) = 3.1392$ £m

expected value (competitive takeover):
$0.4 (17.8479 - 15) = 1.1392$ £m

expected value of option 1: $3.1392 + 1.1392 = 4.2784$ £m

Option 2: retain Bodmin brewery
expected value: $0.5 (8.9059 + 5.6464) = 7.2762$

expected value (smooth takeover):
$0.4 (7.2762 - 10) = -1.0895$

expected value (competitive takeover):
$0.4 (7.2762 - 15) = -3.0895$

expected value of option 2: $-1.0895 - 3.0895 = -4.1790$ £m.

Thus, option 2 with its negative value looks to be a non-starter compared to option 1.

The probability of the bid failing is $1.0 - (0.4 + 0.4) = 0.2$. In the event of such a failure, the expected value of the capital gain of £2m on Redwood shares is $0.2 \times 2.0 = 0.4$£m.

Thus, the overall EV of the Midland bid strategy is:
$(4.2784 + 0.4) = 4.6784$ £m.

The Axis analyst bounced these preliminary figures off Alan Walker and his colleagues for general reactions. He showed them the results of 'sensitivity analysis'. For example, all the previous computations had assumed a hurdle rate of 15%. What if interest rates rocketed? Using his computer spreadsheet, he showed the effects of various changes in basic assumptions about the model. He also wanted to get some rule-of thumb indications of how they would rate the three strategies on qualitative aspects. He asked them to score the following from 1 to 10 (1 = bad; 10 = good):

- fit with consumer taste
- pay back period (how many years before the PV sum exceeds the original investment?)
- fit with Redwood identity (borrowing history, brewing history, traditions, culture etc.)
- fit with Redwood's desired image (solid local employer, traditional but with dynamic edge, change to meet opportunities not for its own sake)
- perceived risk (probability of damaging outcome, effects of rise in interest rates, consequences of poor return)

Table 10.1 summarises the client set's qualitative scores.

Table 10.1

Strategy	Consumer demand	Pay back	Identity	Image	Perceived risk	Total score
A: Dartstone	5	3	6	6	5	25
B: fast food	7	7	3	5	4	26
C: Midland	6	3	5	8	4	26

For the final evaluation, the consultant drew up a table (Table 10.2) for comparing each option.

Table 10.2

Strategy	Assessment measure	
	Quantitative EV	Qualitative total score
Buy Dartstone option 1: retain brewery	−1.5205	25
option 2: sell brewery	−0.2798	
Fast Foods option 1:	2.6764	26
option 2:	1.1927	
Buy Midland option 1: sell Bodmin	4.2784	26
option 2: retain Bodmin	−4.1790	

4.6784 combined

Step 8: Making a choice

The Axis analyst reported back once again to Alan Walker and his colleagues with his findings and the above summary assessment table. Although the overall qualitative score for the Dartstone option was roughly the same as for the other two, on financial grounds this acquisition would result in a loss even if Dartstone's brewery were sold. Strategy A was eliminated.

The fast food strategy is a strong contender financially. The investment required is a lot less and the pay-back period shorter than for the Midland acquisition. However, such a move would take Redwood into unfamiliar territory with commercial hazards and risks they know little about. Fast food represented an uncomfortable fit with Redwood's identity and desired image.

On purely financial grounds, the best bet appeared to be to buy Midland Breweries and then sell their own Bodmin brewery. Such a move would also be well in keeping with Redwood's world-view, traditions and expertise. It would take them from minor to middle league in the brewing industry. Once this venture had settled, Redwood could always reconsider a fast food operation.

Thus, the choice was not made on a 'bottom line' basis alone, although clearly financial considerations are very important. In the event of, say, the fast food strategy indicating better expected values than the Midland strategy, there would be nothing to stop the client set still choosing Midland simply because it fits their world-view and instincts more closely.

Step 9: Implementation

Once the Midland strategy had been decided, Alan Walker set about the task of raising finance to see the acquisition through. His task was greatly helped by the facts, figures and general understanding gained during the three months Axis study.

10.3 Datatrieve – a study of customised software design

This case concerns computerisation of the information processing needs of one of Tony Boyle's clients. The client had noted that their information processing needs were growing rapidly and meeting them had become a problem. Tony was called in as a consultant to examine, among other things, whether computerisation would provide a more efficient information system than the existing manual one. From here on, we refer to the consultant as the analyst in keeping with the rest of this book. The analyst adopted a structured three-stage approach of system analysis, design and implementation.

Step 1: Groundwork

The client set comprised two people both of whom were experts in occupational health and safety. Although they had limited knowledge of computers, they were in many respects 'ideal' clients. They were very self-motivated and enthusiastic and treated the project as a learning experience from which they could draw benefit for future developments.

The client set's world-view was that information was a powerful resource that could and should be harnessed to help them prevent accidents and improve the health and safety of the workforce. This world-view was shared by the client organisation whose board had approved the project and authorised its budget. The analyst's own world-view was also very much in accord with the client set's.

Communications and mutual confidence were established through a series of meetings at which the analyst discussed his approach to the project, what he intended to do and what would be expected of them. His terms of reference and the broad aims of the study were discussed and confirmed.

Step 2: Awareness and understanding

The first task was to break the information processing requirements into sensible 'chunks' and to examine the case for using computers for each chunk. In view of space limitations, we have only covered one of the chunks here – the Datatrieve program.

In practice, Datatrieve was a very small part of the client set's total information processing requirements, the majority of which were met by purchasing 'off-the-shelf' computer programs. For example, all their accident data recording and analysis work was carried out using a commercially available software package designed and written specifically for this purpose. Similarly, they were able to keep track of periodic inspections and monitoring by using the diary facilities of commercially available software.

As mentioned above, the client set wanted to learn more about how computer programs were developed with a view to developing their own database systems in the future. Since Datatrieve was a comparatively small and non-urgent part of their total requirements, it was an ideal candidate for development in which they could participate and learn. The analyst could also approach the project in a less formal way than would be the case in a larger project.

The client set had a large and growing number of books, articles and newspaper clippings and other kinds of print-based material. These had been collected over a number of years and retained for possible future reference. They were filed according to storage location using a simple alphanumeric code (letters and numbers). For example, A1 meant 'first file on top shelf, left-hand end'.

The client set's problem was that although they knew they had useful information, and might even remember a particular document, they had no sure way of quickly locating it. Some searches might require a complete sweep of every file in sequence until the document was retrieved. With the quantity of documents growing, they realised that the retrieval problem would get worse.

The client set had a number of ideas about how the system could be improved such as creating a bibliographic database for identifying each document. However, the analyst advised that rather than them searching for a solution it would be better for them to specify what they wanted to get out of the information system. System description as outlined in Chapter 2 was inappropriate here. What was needed was the Input-Output or 'Black Box' technique.

Applying the Black Box technique

In theory, any information system can be completely specified by describing the outputs required from its use and the inputs that would be needed to produce such outputs. What goes on in between, i.e. the abstract system of information processing, must process inputs in such a way that the desired outputs occur. Details can be ignored as far as system design is concerned.

The 'Black Box' technique is used widely by analyst-programmers. They can get on with designing the system inside the notional 'Black Box' and the clients only have to consider the inputs and outputs.

Although the 'Black Box' technique is very useful in designing information systems, you should use it elsewhere with caution. As Fig. 1.1 in Chapter 1 and our general attitude throughout the book shows, we have deliberately discouraged the 'Black Box' approach. This was to ensure that you did not get into the sterile habit of thinking of *all* systems as 'black boxes'. As an experienced systems person, you can afford to switch your thinking between detailed systems description and 'black boxes' as you see fit.

Step 3: Objectives and constraints

Locating A TOPIC

After discussion and checking, the analyst summarised the position that the client set wanted to reach as: 'If I have a problem with, for example, A TOPIC, I want to enter the information system and quickly find all the alphanumeric locations (i.e. files) which contain information on A TOPIC.'

The main objective was therefore straightforward. However, there were a number of constraints to take into account. These were categorised as 'selection', 'resource' and 'technical' constraints.

Selection constraints

A TOPIC could not be anything in the world! The range had to be limited to topics likely to be relevant to the client-set's area of work, i.e. relevant to occupational health and safety. Fortunately, the analyst was able to rely on a published list of topics that covered the desired range.

Resource constraints

As noted earlier, Datatrieve formed only a part of the client set's information processing problem. The overall project had a fixed budget and time for completion but the analyst was given considerable latitude on completing Datatrieve. It was regarded more as a development project.

Manpower and time were identified as major constraints in coding and sorting all the existing documents in the client set's 'library'. It was decided to restrict coding and entry of data into Datatrieve initially to a selection of existing documents that could be used to test the new system. Any new documents would also be coded and added to Datatrieve.

Technical constraints

A computerised system was not a foregone conclusion but would clearly be a main contender. The client organisation had a policy on the computer hardware and software that could be used within the company. Essentially, any program had to run on an IBM–AT with 640K of RAM (user memory) with MS-DOS 3.3 as the operating system. Under the company's computer policy, this was the only equipment available to them. The company also had a policy that any database applications such as Datatrieve should be written in dBase III, a particular computer language designed specifically for this kind of application. It is quite common for client companies through their computing departments to specify computer languages since programs may require revision in the future and the company's own programmers will have to understand the language used. By placing such constraints on software development, the client company has the option of undertaking their own revision if, for example, the original software authors charge very high fees or are no longer available.

Analyst's objectives

The analyst was aware of the client set's objectives and constraints. His own overall objective was to produce within budget and within agreed timescales an error-free Datatrieve information system that would meet the client set's project objectives. He drew up an objectives hierarchy including such sub-objectives as 'specify inputs in detail', and for possible computer options 'design screen layouts', 'draw up detailed program spec', 'write prototype program', 'test prototype' and so on.

Step 4: Developing Datatrieve

Analysis of options

There are a number of strategies for developing an information system capable of producing desired outputs. One way is to improve the existing manual system. Among computer options, usually, the cheapest and easiest is to buy an 'off-the-shelf' program and this is satisfactory for most general-purpose applications although it would be limited for unusual tasks. Another route is to buy a ready-made program and have parts of it tailored to meet your particular requirements. This is more expensive than the off-the-shelf route and not all programs enable tailoring to be done. The third computer option is to have a program custom-written for your precise require-ments. This is the most expensive but usually the most effective computer route. In a first quick pass through the analysis phase (not covered here), a structured systems method similar in principle to SSADM was used to critically examine these options.

The clients established before the analyst was approached that their current manual system could not be improved enough to meet their long-term requirements. The absence of an index and the cost of introducing and maintaining one had weighed heavily in their decision. The analyst could have spent some time formally analysing the possibility of upgrading the manual system. However, in view of the client set's determination to give up the manual system, he relied on his own knowledge and experience to judge for himself that an improved manual option would not meet their requirements.

Investigation of off-the-shelf and tailored programs showed that at least one of these (Cardbox) could satisfy most of the requirements so far as straightforward information retrieval was concerned. However, using such an off-the-shelf package would meet none of the client set's needs for learning about the design and development of software. If they were restricted to an off-the-shelf package, they would have little say in how the program would 'look and feel'.

In investigating off-the-shelf and tailored options, the analyst had to keep in mind the fact that the client set were not experienced computer users who were knowledgeable about the intricacies – and potential problems – of using such methods. For example, a computer specialist having the client set's requirements might try a much more sophisticated approach using an OCR (optical character recognition) scanner to capture the current manual records, an off-the-shelf word processing package for text editing, and an off-the-shelf text retrieval package. As the Tech-Abs case in Chapter 4 shows, OCR scanners linked to text retrieval systems can be fraught with technical problems. In the hands of naive users, such an approach might well create a bigger problem than the one it was intended to solve – the counter-intuitive outcome.

For these reasons, only the custom-written Datatrieve option was taken further. There is little economic sense in trying to develop all four options in parallel. Thus, once a structured systems analysis had eliminated other options, Datatrieve represented iterative development of a single solution, i.e. customised software. The routes that were modelled and evaluated in later stages were thus different variations of the Datatrieve program developed and tested iteratively as described below.

From analysis to design

The creation of possible routes was broken into two phases, both of which recognised the client set's lack of computer expertise. In the first phase, relevant examples of other computer programs were demonstrated and principles underlying this sort of information processing were explained. This enabled the client set to take an active part in creating a number of different possible solutions to their problem. As this phase proceeded, the client set also saw further ways in which Datatrieve could help them with related problems. However, for simplicity we will stay with Datatrieve's main function – entering and retrieving information.

In phase two, it was decided to work iteratively with the client set towards the best solution for them. On the basis of possible solutions identified in the first phase, a range of programs could have been written, each one emphasising a particular valued feature such as graphical displays, browsing and so on. However, expecting the client set to make an informed choice between such options would have been unrealistic, given their limited computer knowledge. Instead, the analyst wrote a simple 'core' program dealing with the main Datatrieve functions. As the client set became familiar with using this, refinements were made and further functions were written into the program.

This iterative kind of approach to program design used to be rare because, once written, programs were difficult to alter. However, modern microcomputers possess a range of facilities that can be used to produce prototypes very quickly.

Step 5: Measures for assessing achievement

The 'acid test' would have to be whether or not the client set felt happy that they could use Datatrieve to solve the information processing problem they had identified. As an assessment measure, 'feeling happy' or satisfied is qualitative but such a measure has a number of quantitative components that contribute to it. For example, the Datatrieve computer file had to be capable of holding at least 10000 topic records. Each record had to be capable of storing a realistic amount of information such as 'author of document', 'date of publication', 'location of document', and 'topics covered'. Just how much information a record might have to contain was a technical problem for the analyst to address. Another quantitative measure was the speed of information retrieval; search times measured in seconds would be acceptable whereas minutes would not.

The analyst had to devise tests in conjunction with the client set that could be used to evaluate Datatrieve later on. For example, several different complex searches for topics were set up so that timings and accuracy of retrieval could be measured.

Step 6: Modelling and evaluation

Although modelling, evaluation and options were developed iteratively, there were two quite separate cycles. In the first cycle, qualitative iterations were carried out by the client set. This was concerned with how Datatrieve 'looked and felt'. Results from this cycle were passed to the analyst who amended the prototype, thereby enabling re-evaluation and refinement of the model.

The other iterative cycle was a quantitative one carried out by the analyst. This involved technical matters such as alternative data structures, efficiency of memory management inside the computer, and efficiency of storing data on disk. The analyst made an early decision to make the two cycles as independent as possible so as to allow the client set freedom to choose how the program 'looked and felt'. The following summarises the first pass plus two iterations.

First pass

The client set decided that Datatrieve should be analogous to a card index so as to give them a conceptual link with a familiar information system. This led to a layout of information on the computer screen as shown in Fig. 10.8. The labels 'author', 'title', 'date of publication', 'topics covered' and 'location of item' are what are known in

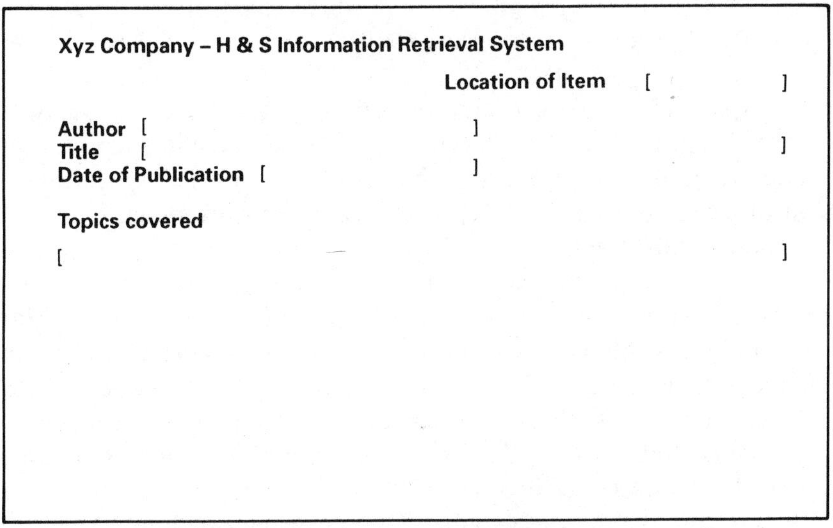

```
+--------------------------------------------------------------+
|                                                              |
|   Xyz Company – H & S Information Retrieval System            |
|                                                              |
|                              Location of Item    [       ]   |
|                                                              |
|   Author  [                    ]                             |
|   Title     [                                            ]   |
|   Date of Publication  [       ]                             |
|                                                              |
|   Topics covered                                             |
|                                                              |
|   [                                                      ]   |
|                                                              |
|                                                              |
|                                                              |
|                                                              |
|                                                              |
+--------------------------------------------------------------+
```

Fig. 10.8 Datatrieve screen layout at first pass.

computing terminology as 'fields'. This layout of headings or field labels is fixed for all the records in Datatrieve and is the same whether a new record is being created (data input) or a completed record is displayed after a search (data retrieval). The blank areas in the brackets next to the labels are filled with items of information relevant to a particular record. For example, Fig. 10.9 shows how a completed record might look using a fictional article on 'asbestos in false ceilings'.

```
Xyz Company – H & S Information Retrieval System

                              Location of Item  [C27              ]

Author [Penny M              ]
Title [ Managing Asbestos in Buildings                           ]
Date of Publication  [March 1988]

Topics covered

[ Asbestos, air monitoring, removal                              ]
```

Fig. 10.9 Screen display of completed Datatrieve record.

To simplify the use of Datatrieve, only one screen layout was provided as in Fig. 10.8. The client could either type in a new record and save it on disk for future retrieval, or type in a search specification in order to find particular types of record. For example, typing in 'asbestos' into 'topics covered' and '1987' into 'date of publication' would retrieve all records concerning asbestos relating to 1987.

Several problems were identified with this screen layout. For example, a single code for location of item was generally accurate enough for journal articles and short documents but too ill-defined for large documents such as books. A TOPIC needed to be locatable by chapter(s) and page(s). Aliases and alternative entries for author name would not automatically be cross-referenced by Datatrieve and so the 'hit rate' on a search was variable. The title field was tedious to type in, was prone to miskeying errors, and the client set felt that in any case it was not very useful. There was no provision for consistent entry of publication dates. The provision of a single field to cover all the topics caused noticeable delays in searching.

First and second iterations

The first iteration sought to remedy problems identified in the first pass. In particular, the title field was deleted, 20 topic fields were introduced, and the number of ways of entering the publication date was reduced to one. For items such as books, an extra location field was added to indicate page numbers.

One of the client set then suggested that Datatrieve could be expanded to include 'anecdotal references' such as useful advice obtained during conversations, snippets picked up at conferences, and so on. In other words, he wanted to upgrade Datatrieve from a database of bibliographic and documented topics to a 'general intelligence' database. This request was fulfilled in a second iteration by including fields for 'notes'. When entering such notes, the client set used similar rules to those used when entering other items except that author referred to the source of the intelligence and date referred to the date of entry. Figure 10.10 shows the screen layout at second iteration using the fictional article on asbestos in ceilings for illustration.

```
Xyz Company – H & S Information Retrieval System

                        Location of Item   [C27                      ]
                        Page Number (s)  [    7]  to  [          25]

Author [Penny M          ]
Title [Managing Asbestos in Buildings                               ]
Date of Publication [      ]          [03]      [1988]

Topics covered

[asbestos    ]   [air sampling]   [              ]   [              ]
[monitoring ]   [fibre counts ]   [              ]   [              ]
[removal     ]   [              ]   [              ]   [              ]
[PPE          ]   [              ]   [              ]   [              ]
[tenting      ]   [              ]   [              ]   [              ]

Notes

[guide to practical procedures with good bibliography              ]
[                                                                    ]
[                                                                    ]
[                                                                    ]
```

Fig. 10.10 Screen display of completed Datatrieve record (second iteration).

Step 7: Implementation

Once the client set were satisfied with the prototype of Datatrieve, and had produced topic lists for all their pre-existing information, the prototype program was converted to its final form. The client set then put it through its paces again to ensure that no 'bugs' or computing errors had crept in during conversion. By this stage, the client set were fully conversant with Datatrieve and so no training was needed.

In general, the complexity of a program increases exponentially with its length. For example, if program B is twice the size of program A it may be four times as complex. Once programs get much bigger than Datatrieve, it is essential to adopt rigorous, formal procedures for testing and prototyping. Without this, fundamental flaws in program design may remain undiscovered until it is too late to remedy them without great expense. The Datatrieve program was small enough to rewrite it completely if it failed to meet the client set's requirements at prototype stage.

The above description of testing and prototyping was from the client set's point of view. In practice, no professional software writer would let loose a program on clients with only this level of informal testing and prototyping. In parallel with the client set's testing, the analyst arranged for another programmer to formally test Datatrieve using a structured set of tests covering, for example, the automatic trapping of errors during data input. Other tests covered data integrity, i.e. minimising the risk of loss of data in the event of a program fault or breakdown. The client set were excused from direct involvement in the tedious, painstaking work of formal testing although they were kept informed of procedures and progress so as to aid their understanding of design and development.

The all-up cost of Datatrieve was £3,000 which was about six times the outlay costs of them simply buying an off-the-shelf data retrieval package and then having it tailored. At first glance, the cost comparison may seem unfavourable but it should be borne in mind that the client set received a two-day formal training course in using their computer effectively plus several days of personal assistance and expert advice during the development of Datatrieve. This approach not only met the client set's objective of learning first hand about the design and development of such systems but it also avoided costly trial-and-error learning that could have arisen with the unaided use of an off-the-shelf package.

10.4 Summary

The two cases studied in this chapter have shown how markedly different kinds of problem may be tackled using hard systems analysis. In the Redwood Breweries case, the application was in the realms of strategic planning and decision making covering a range of uncertainties about current and future markets, business trends and strategic choices. Here, hard systems analysis enabled uncertainties to be rationalised and options to be clarified by quantification, i.e. a fog-clearing exercise.

The Datatrieve case, however, was at a tactical level and showed hard systems analysis applied to development of an information system. The initial phase of structured analysis enabled a decision to be made between a number of competing options, including improved manual and computerised approaches. Once customised software had been chosen, the second phase was the development of that software. This was an example of systems engineering which, in this case, focussed on solutions early in the application, i.e. even though aspects of a structured methodology were used it had many similarities with formal problem solving as described in Chapter 9. With Datatrieve, the model became the solution and, rather than a close examination of several potential solutions all the way through, iterations concentrated on refinement. Not all systems engineering cases are like Datatrieve. There are a number of variants of systems engineering methods and some suitable references are given in Useful Reading at the end of this book.

Chapter 11
Soft Systems Analysis

11.1 Introduction

Chapter 6 introduced the soft systems method (SSM), otherwise known as the Checkland method. In this chapter, we review the method and examine in more detail some of its trickier steps, in particular the CATWOE test and conceptual modelling. We suggest that you turn to Fig. 6.1 in chapter 6 now to remind yourself of the SSM steps.

11.2 Step 1: Data collection

Collecting information about the 'mess' follows the same kind of approach as data collection in the hard and failures methods. Interviewing key figures is usually a necessary part of the process but you need to pay particular attention to expressions of dissatisfaction and concern. Equally, watch out for signs that a key figure apparently is satisfied or unconcerned whereas other key figures are not. Often apparent lack of concern about a problem situation may be contributing to the mess.

Exercises

1. What is the systems definition of a 'mess' as given in Part 1?

There are no hard and fast rules for deciding who are key figures. Typically, the client or problem owner will provide you with background information and may offer you an opinion as to what is wrong. Certain individuals may be cited as being involved and these can be assumed to be key figures. However, as your interview programme develops, further key figures may be identified. How many you interview depends on your resources and time constraints.

Interviews should be non-directive. Ask open-ended questions that start with phrases such as 'Tell me about ...', 'Why ...?', 'What ...?' and so on. Avoid leading questions that presume what the answers will be. Interviewees often expect some kind of instant diagnosis, and commentary on the situation should be kept as little and as neutral as possible. Your views should only be revealed at steps 5 and 6 of the method. All interviews and your notes on them must remain confidential; this is a basic tenet of all consultancy and research.

Documentary information may also be useful. Typical examples are minutes of meetings, reports, memoranda and correspondence. To a large extent, you have to rely on what key figures consider to be important information. Sometimes documents that would be invaluable to your task are considered too sensitive to be revealed. You just have to accept this inconvenience unless the client considers that all relevant documents are to be made available to you. This would be one of the negotiating points when agreeing to do the soft systems study.

At the end of data collection, you should have available sufficient information about the unstructured problem situation to enable you to proceed to analysis.

Exercises

2. What clues to the existence of a soft systems problem would you look for in the collected information?

11.3 Step 2: Analysis

Analysis begins by drawing a rich picture of the messy situation. A rich picture is the analyst's own interpretive 'snapshot' of the mess. Remember, a rich picture is not a system diagram. Although you are allowed to put loose boundaries around things for ease of understanding, such boundaries should not attempt to represent systems or subsystems.

The technique of constructing a rich picture is described in Chapter 5. There are several examples of rich pictures in Chapters 2, 5, 12 and 14.

From the rich picture, you should be able to identify a number of issues and perhaps one or two primary task areas that seem important to the situation. Clues about these are to be seen in symbols and expressions of clashes, pressure and uncertainty.

Problems in the Health Service – a mini case study

Several attempts to reorganise the Health Service in the United Kingdom have been made in recent years. Since 1974, a three-tier organisation of Regional, Area and District health bodies has operated. At each level, multi-disciplinary teams are supposed to function so as to achieve cohesion and a unity of purpose in both principle and practice. This intention has not always been met. At District and hospital levels, for example, it is not uncommon today to find numerous value clashes: nurses vs. doctors, clinical staff vs. admin staff, consultants vs. everybody, everybody vs. 'Area, Region, and the Department' (of Health). Although health care is the assumed and often stated common purpose of all these factions, it is apparent that they often pursue different and conflicting objectives. In any event, health officials at almost every level complain about poor performance. A particularly graphic example is the account of the Normansfield Hospital Crisis in 1976 given by Victor Bignell and Joyce Fortune in 'Understanding Systems Failures' (1984). The crisis in question was a strike by nursing staff that was 'unprecedented in the history of the National Health Service'. The account reveals a classic soft systems problem with an escalatory spiral of conflicting values, objectives, clashes and communications problems that led ultimately to organisational failure – the strike. More recently, examples have been cited of particular health authorities in which 'health service officials are resigning and swarming to leave' and managers being 'broken and suffering from battle fatigue'.

From a rich picture of the above, you might derive the following issues and 'problems' related to the primary task of providing patient care:

Issues	*Primary task-related areas*
leadership	hospital administration
decision-making processes	clinical support
morale and motivation	ancillary services
authority and power	
managerial competence	
industrial relations	
coping with change	

11.4 Step 3: Relevant systems and root definitions

As outlined in Chapter 6, steps 1 and 2 deal with what you perceive to be happening in the real world – 'what is'. Step 3 requires a complete shift of thinking to the consideration of hypothetical systems.

The first task in this step is to dream up relevant systems for your list of issues and primary task areas. In the previous Health Service example, there are probably too many issues and primary task areas to carry; as some issues are clearly related, you could choose to coalesce some of them, for example. Aim for three or four issues and one or two primary task areas.

Who decides what is relevant? Ultimately it will be the actors when you debate with them in step 6. However, in step 3 you must use your own judgement. As noted in Chapter 6, you need to be subtle in naming your relevant systems. In particular, you need to deliberately avoid focussing on 'obvious' input–process–output ideas as these are likely to fix your thoughts back in the real world. To quote the Lucrative case from Chapter 6, 'a system for taking editorial copy plus advertisements and converting them into marketable yearbooks' is so real world and unimaginative that its value in stimulating insights is virtually nil.

In the Lucrative case in Chapter 6, we teased out two relevant systems which we will now call RS1 and RS2. RS1 started as 'an editorial and advertising reconciling system'. Such a title is a short-hand description and it needs to be defined so that it makes a bit more sense. Our first go at the Root Definition (RD1) of RS1 was 'a system to ensure that editorial decisions reflect the best interests of the publishing house'.

A root definition needs to be adequate otherwise you may find difficulty with later steps in the method. There are two complementary ways of testing whether a root definition is adequate. The first is to assess it critically by examining it for ambiguities, woolliness and implicit assumptions that may be questionable. The result of this first iteration of RD1 was 'a system to be operated by Lucrative's managing director to ensure that editorial decisions reflect the objectives of the publishing house in terms of financial viability and profitability and in terms of reputation in the marketplace among purchasers and advertisers'. RS1 was also modified to 'a system for making cost-effective editorial decisions'.

The second test is more structured – the CATWOE test. CATWOE stands for:

Customers
Actors
Transformation
Weltanschauung(en)
Owner(s)
Environment

Exercises

3. Define the terms that make up the CATWOE mnemonic.

If you can clearly and unambiguously identify the CATWOE elements in your root definition, then it has passed the test with flying colours. However, do not worry if you do not achieve this paragon. What is important is that you check and probe the root definition and if necessary justify to yourself why a CATWOE element is absent.

Applying the CATWOE test to the Lucrative RD1 (first iteration):

- *Customers*: purchasers and advertisers explicit; editorial staff, MD implicit.
- *Actors*: Managing Director explicit, although would he really be operating such a system? other staff implicit.
- *Transformation*: editorial decision making (see Fig. 6.2 in Chapter 6); the process is *not* 'converting editorial copy plus advertisements into marketable yearbooks'.
- *Weltanschauungen*: there is really only one world-view that is relevant here – the view that editorial decisions have to take account of economic circumstances and satisfice the legitimate interests of the system's customers.
- *Owner*: the Managing Director is both system owner and owner in title.
- *Environment*: RD1 does not explicitly state constraints but some obvious ones are sources of finance, competitors, bank lending rates, attitudes and changing tastes of purchasers, advertiser behaviour.

The second relevant system (RS2) in the Lucrative case was 'a system to satisfy the needs of customers'. After iteration, this was modified to 'a system to satisfy the needs of yearbook purchasers' with a root definition (RD2) thus: 'a system owned by Lucrative which aims to satisfy the needs of purchasers of its range of books coupled with a return on investment commensurate with at least maintaining its share of the yearbook/handbook market.'

Exercises

4. Critically assess RD2 and apply the CATWOE test to it.

As a result of critical examination, it is now possible to revise RD2 thus: 'a system owned by the Managing Director and operated by all Lucrative's staff which seeks to satisfy cost-effectively defined needs of purchasers of its range of books and so at least maintain its defined share of the yearbook/handbook market and thereby secure for Lucrative a defined adequate return on investment taking into account economic, product market and labour market constraints'. RD2 is improving but doubtless it needs further iterative development. The title of RS2 can now be revised to 'a system to satisfy cost-effectively the needs of yearbook purchasers'.

11.5 Step 4: Conceptual modelling

A conceptual model for RD1 was developed in Chapter 6 (see Figs 6.2 and 6.3). Remember that a conceptual model depicts only what the system *logically* would have to comprise in order for it to function.

Figure 11.1 shows a first attempt at a conceptual model of RD2. Note the level of abstraction in the main activities. There is no mention of real-world practical things like 'set up sales order system on computer' or 'print yearbooks by web offset litho' or 'send mail shot letters to potential purchasers'. Note also that the number of main activity verbs has been limited to seven deemed to be essential in this case.

Testing conceptual models

In a similar way that root definitions have to be tested, so too do conceptual models. Tests based on the formal system model (see Part 1, especially Chapter 8) include the following adapted from Peter

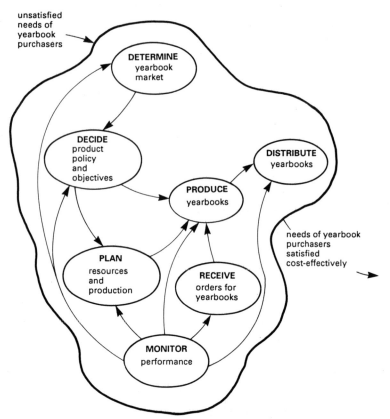

Fig. 11.1 A conceptual model of a system for satisfying cost-effectively the needs of yearbook purchasers (first iteration).

Checkland (1981) which we have applied to RD1 and Fig. 6.3 (comments in brackets):

(1) Does the model and its root definition suggest a continuous and relevant mission? (Cost-effective editorial decisions are obviously needed as a permanent feature of a publishing house.)
(2) Can performance be measured? (The model includes the establishment of criteria and monitoring and control activities).
(3) Is a decision-making activity present? (Yes.)
(4) Do any of the main activities comprise sub-systems of back-up activities? (Fairly obviously they do. For example, in order to determine market needs and wants one has to:

   identify: markets
   research: markets
   assess: research data.)

(5) Do the system components interconnect? (Yes.)
(6) Does the system interact with an environment? (Assume, for example, cash limits, market size, trades unions, legislation, space limits etc)
(7) Does the system have a boundary? (The wider system here is the Managing Director. Although he owns the system and may be involved in editorial decisions from time-to-time, the system is essentially operated by other actors, e.g. editorial, advertising, finance staff.)
(8) Can the wider system provide resources? (Assume that the Managing Director can appoint competent and sufficient staff and provide working accommodation, equipment, stationery, etc.)
(9) Can the system be sustained? [The Managing Director as system owner (and actual owner of the company) is a sufficient power figure to ensure continuity of the system if he wishes it.]

Activity

Carry out the test on the conceptual model of RD2 (Fig. 11.1) and iteratively modify it as appropriate.

Note in test number 4 above the reference to back-up activities. Identifying these is an important part of expanding a conceptual model. You carry on expanding, testing and modifying until in your judgement you have sufficiently clarified what such a hypothetical system would have to comprise for it to function. Figure 11.2 depicts an expanded version of the conceptual model of RD1.

Activity

Expand your conceptual model of RD2 so that it incorporates back-up activities and then test it using the nine-point check-list.

11.6 Step 5: Comparisons to provide debating agenda

As you will note from Fig. 6.1 in Chapter 6, step 5 returns to the real world but you must still avoid thinking or expressing yourself in terms of real-world practical *solutions*. The aim of the comparison step is to

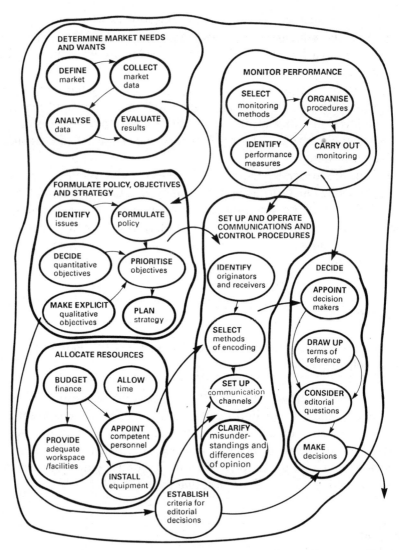

Fig. 11.2 Expanded conceptual model of a system for making cost-effective editorial decisions (first iteration) (see Fig. 6.3).

illuminate the problem situation analysed in step 2 both for yourself and the actors and not to state or suggest that *you* now know *what* must be done.

In step 5, you are comparing 'what might be' (conceptual model) with 'what is' (real-world situation from step 2 analysis). There is no single best way of carrying out the comparison. Each analyst develops his or her own approach and some are more formal than others. One way is to use the conceptual model as a reference and check whether each activity occurs in the real world situation. If it does, why does it? who carries it out? has it always been done? why is it done like it is? and so on. Similarly, if it does not occur why is this so? This approach tends to illuminate activities whose existence or absence in the real world setting is taken for granted and therefore whose potential contribution to the problem situation may not be recognised. This approach may thus be especially useful with issue-based problems.

A second approach is to imagine the conceptual model in operation in the real world and note all the practical implications. Then consider how the same task is actually carried out in the real world. Mismatches and discrepancies should raise important questions. This approach is well suited to problems embedded in primary tasks.

A third approach uses a template method. A conceptual model of what actually happens in the real world is constructed from the rich picture. The activities are derived to follow as closely as possible those of the conceptual model from step 4. By overlaying one model on the other, it should highlight matches and mismatches and these can be tabulated systematically for ease of reference.

Whichever approach you adopt, out of it should come an agenda for debate with the actors. This agenda is a list of topics covering mismatches, omissions, etc., together with 'what?' and 'why?' questions to stimulate the actors' thinking. The agenda still needs to avoid real-world prescriptive solutions.

It may also be useful to be able to show the actors a rich picture of the current situation that is perplexing them. However, if your rich pictures are, like ours, a little on the rich side, it would be wise to draw up a 'sanitised' version of your rich picture from step 2 for such public consumption. This is a good general rule for rich pictures used in any of the methods as there is no value in causing offence to actors who may appear in the rich picture in an unfavourable or ridiculous light. Whether or not you show any rich picture at all is a matter of judgement. Will they be offended? Will they regard rich pictures as 'expensive doodling'? Will you go up or down in their esteem? Our advice is, if in doubt keep them to yourself.

11.7 Step 6: Discussing the agenda with actors

For even experienced analysts, this stage can be the most anxious. The discussion takes the form of a debate with those in key roles such as client, problem owner and actors identified in the CATWOE test. The important words here are 'discussion' and 'debate'. Analysts used to acting as conventional management consultants tend to act out the role of solution provider. In such a role, the analyst possesses 'expert power' grounded in the shared belief that the expert's task is to provide a prescriptive solution to the problem. In SSM, however, the analyst's role is more that of a *therapist*; he or she may well possess considerable expertise and may well be capable of recommending solutions, but such a temptation must be resisted. A debate does not consist of the analyst speaking all the words of wisdom and supine actors soaking it all up.

Nevertheless, actors on the client side may attempt to seduce the analyst into giving his or her solution. This is likely to happen in organisations in which collective rather than individual responsibility is the tradition or in which 'fighting one's corner and watching one's back' makes buck-passing an endemic disease. Individual actors may feel very uncomfortable with having to address issues on the analyst's agenda for debate and may prefer to transfer the mental effort back to the analyst. The analyst should be prepared for the full range of manoeuvring and attempts to impose control such as: 'Well *you're* the expert. What's *your* answer?' You, as analyst, are not there to provide practical solutions but to get the actors to examine critically your conceptual model of a possible future system, i.e. to agree a change from 'what is' to 'what might be'.

You need to try as far as possible to organise and express the agenda in such a way that your ideas for change (stemming from rich picture, root definition, conceptual model and comparison) are not only desirable in system terms but are also culturally consistent. Systemic desirability is relatively easy to address since much of your work on root definition and modelling will have focussed on the logical requirements of a system. The debate needs to check whether what you are suggesting is deficient or contains logical errors.

Aspects of cultural feasibility have been raised in previous paragraphs in so far as they affect the course of the debate itself. Chapter 6 discussed some of the difficulties surrounding the cultural aspects of proposed changes. But, whereas culture is an easy label to apply, as a phenomenon it is very nebulous and difficult for the analyst to handle. For culture is not a 'thing' but a complex and dynamic property of human activity systems; it encompasses unwritten and usually unadmitted rules of behaviour, ideologies, habitual responses, language, rituals and so on. Although one can refer to organisational culture as if it were a uniform 'thing', in fact organisations include micro-cultures; different functions and specialisms have their own tribal interests, values, language, attitudes, idiosyncrasies etc. that characterise them. Remember the characteristics of the Health Service in the mini-case study earlier in the chapter? The actors with whom you debate will share features not only of the organisational culture but individual actors will represent particular micro-cultures. Cultures serve to protect members' interests and tend to be slow to change; cultural change tends to be evolutionary rather than rapid (see also Chapters 16 and 17).

You should anticipate therefore that even if actors agree with the systemic desirability of your proposed changes, they might oppose some or all of them on what may be loosely described as 'cultural grounds'. If and when this occurs, you just have to accept the actors' verdict however disappointed you may feel. Remember, a therapist's task is not to 'cure the patient' but to help the patient face up to what may be ailing him and to choose whether or not to work towards a better state. We do not agree with management consultants, especially in the information technology field, who offer off-the-shelf 'cultural change packages' designed to facilitate the introduction of new technology. Organisational culture is not a disease to be cured and, if history tells us anything, cultures are resistant to rapid changes whatever the promises of the latest prescription. If your proposed changes involve changes in cultural aspects, all you can do is sow the seeds in the actors' minds and bring to their attention the slow nature of such changes.

As outlined in Chapter 6, agreed changes typically will fall under the headings of structural, procedural, policy and 'attitudinal' (or cultural). If none of your proposed changes relating to a root definition are accepted for whatever reason, you can switch to one or more of your other root definitions. It is unlikely that all your proposed changes relating to, say, three root definitions will be rejected after debate. If you have only drawn out one root definition (because of lack of time, perhaps), you can go back to the rich picture and identify some more relevant systems for working up and subsequently another debate with the actors.

11.8 Step 7: Action for change

The actors should now possess a hit list of agreed changes expressed in terms of 'whats' rather than 'hows'. Converting whats into hows could involve formal problem solving if the options are clear or could suggest hard systems analysis for complex matters (as described in Chapter 9). In principle, the soft systems analyst's task is complete at the end of step 6 but we recommend that you offer continuing support to the actors. Even if the actors agree a number of proposed changes and appear to accept their cultural feasibility, hidden agendas may surface after agreement that serve to thwart implementation of the changes. The actors may simply not know how to proceed with practical implementation.

11.9 Summary

The SSM offers a rational tool for tackling problem situations relating to human activity systems that are wicked, messy, chaotic or seemingly intractable. Rather than attempting to define 'the problem' and then solve it, SSM identifies issues and primary task areas from which hypothetical systems and conceptual models may be derived. Comparison of these logically defensible systems with the real-world situation provides the basis of debate with the actors, in essence a learning process for them. SSM assumes that the actors are totally rational and are not subject to personal or cultural influences that may deter them from either agreeing proposed changes or implementing them once agreed. Chapter 12 provides two worked case studies of the SSM.

11.10 Suggested answers to exercises

1. A mess is defined as a system of problems that defies resolution simply by trying to solve each problem one-by-one. The messiness and intractability are emergent properties of the system of problems, i.e. they are additional to the component problems.

2. Evidence of a soft problem would include: references to 'a mess', 'shambles', 'worry', 'concern', 'dissatisfaction', 'dispute', and signs of anger, resentment, mistrust and seemingly inexplicable poor performance.

3. Customers are those affected by the system; not to be confused with customers of the company, although sometimes they may be.

   Actors are the social actors who operate the system, i.e. have roles to play in it; mostly they are role categories and groups rather than individuals.

   Transformation represents the system's essential process and is usually quite different from real world practical processes.

   Weltanschauungen are the world-views of actors; usually this is taken as a shared world-view that would be relevant to the hypothetical system and the organisational culture in which it resides.

   Owners are power figures who control the system and allow its very existence; they are not necessarily owners of the organisation.

   Environment represents external constraints and influences.

4. The main problem with RD2 as it stands is that it has three aims: satisfying purchasers, securing return on investment, and maintaining market share. Each of these targets is open to various interpretations. What does 'satisfying purchasers' mean exactly? Getting books to them on time? Reducing complaints to a minimum? Responding to readers' suggestions for new sections? Is not return on investment dependent on market share?

   Applying the CATWOE test to RD2:

   Customers: purchasers explicit; but also Lucrative.
   Actors: explicitly all Lucrative's staff.
   Transformation: taking unsatisfied needs of purchasers and satisfying them cost-effectively.
   Weltanschauung: the view that purchasers' needs have to be met but within the bounds of economic prudence.
   Owner: explicitly Lucrative's Managing Director.
   Environment: financial resources, interest rates, competitors, labour market (experienced, competent staff may be hard to recruit), advertiser behaviour.

Chapter 12
Soft Systems Case Studies

12.1 Introduction

Chapters 5, 6 and 11 laid the groundwork for understanding and using the Checkland or Soft Systems Method (SSM). In this chapter, we present two case studies that demonstrate the practical application of SSM. Both cases concern academic organisations but you should not infer any significance from that fact or conclude that they will only be of interest to academics. The first case is that of a publishing project that got into a crisis involving finance, staff relations and project management. The second case concerns a long grumbling sense of dissatisfaction among the teaching staff of an environmental health department of a polytechnic. Both cases highlight the relevance of micro-cultures and value clashes to soft systems problem situations.

12.2 Northbrook Training Materials Unit – a study of a project in crisis

Step 1: Data collection

The following is a summary of data collected by the external systems consultant from interviews with staff at the Training Materials Unit, minutes of progress meetings and other relevant documentation.

The year 1983 saw the start of a major training initiative by the Manpower Services Commission (MSC), a government department charged in part with addressing skill shortages in industry and commerce. The MSC and commercial sponsors together set aside multi-million pound 'pump priming' funds to help redress the skill shortage via open learning. Open learning entails the use of courses and learning materials that are made available with very few restrictions in the entry requirements. Typically, open learning materials comprise course workbooks, videos, audio-tapes, and assessments. This practical systems book for example, is written very much in open learning style.

A large part of the sponsored fund was available for bids by production and/or 'delivery' projects. Many such projects sprang up within existing bona fide educational or training organisations. The Training Materials Unit at Northbrook College was such an example.

Northbrook's Training Materials Project soon found itself in possession of £600,000 'pump priming' money to cover an initial three-year period, at the end of which the project would have to be self-financing.

Northbrook College is by tradition dominated by engineering and technology faculties. Malcolm was an ambitious Senior Lecturer in the engineering faculty and the advent of the Open Learning initiative provided him with an opportunity to advance his career ambitions. The College's Directorate were dyed-in-the-wool traditionalists when it came to education and did not know the first thing about open learning. Nevertheless, they appreciated the possibility of gaining kudos and 'Brownie points' in the increasingly competitive world of further and higher education; if the sponsors were happy to pay for the experiment, what had the College to lose? Malcolm was allowed to make his bid which was successful, but like all grants there were strings attached. The money could only be used for specified purposes and had to be accounted for. Crucially, the money was intended only to fund the initial development and production phase, the assumption being that sales of the materials would then make the project self-financing within three years.

Malcolm had two great strengths. First, he was a visionary who foresaw the potential of open learning in meeting needs of employers and employees during the rapid structural changes of employment and jobs that were gathering momentum. Second, he was good at 'stand up' expositions of his case; he was able to convince College and sponsor officials, at least in the first two years of the project, that everything was going quite smoothly. He was similarly able to convince recruits to the project that they had a golden future ahead of them.

However, Malcolm also had a number of weaknesses. At a practical level, he had no experience of managing large sums of money or business projects with deadlines attached. Lack of management experience began to show itself all too clearly as production deadlines came and went without sign of real progress. To be sure, Malcolm held a 'progress meeting' with the project team once a week but as month after month went by it became clear that what was being monitored was 'slippage' rather than progress. As one of the project team cynically noted, 'We ought to call these Lack-of-Progress Meetings.' Malcolm's personality became an added dimension when it became clear to members of the project team that 'things' were going wrong.

The recruitment plan was to hire a mixture of academics (i.e. college lecturers) and 'outsiders' who had experience of writing or publishing academic materials. As it turned out, the three outsiders Joan, Jim and Don also had adult teaching experience which was an advantage.

Joan had substantial publishing experience as an editor and was ideally suited to manage the production side. However, she was appointed on an administrative grade whereas the other team members, whose main task was materials design and authoring, were appointed as academic staff members. This distinction was to have unfortunate consequences.

Although Jim had teaching experience in higher education, he was more commercially minded than most of the team. Early in the project he was concerned at the lack of market research and wondered whether Malcolm's assumptions about who would buy the course materials and in what quantities were warranted. Nevertheless, like Joan and Don, he was persuaded for the first year that Malcolm, the College and the sponsors must know what they're doing.

Don was the most experienced in writing open learning materials. Like Jim, he was returning to the academic world but he was much more of a purist. Whereas Joan and Jim wanted to temper academic excellence with economic prudence, Don was more interested in getting things right. Nevertheless, Joan and Jim were somewhat in awe of Don's expertise and he often won his case for an extension to his deadlines.

Avril, Bob and Susan had no previous experience of either professional writing or open learning. However, they had solid teaching experience in further education and it was planned that they would 'learn the ropes' under the watchful eyes of Joan, Jim and Don. In the first year, however, Avril and Co. took a lot longer to plan and start to write their learning materials than expected. Allowances were made for inexperience and Don encouraged them to put accuracy and polish before meeting deadlines.

By mid-1985, storm clouds were gathering. Joan and Jim were losing patience with Malcolm for failing to inject a sense of urgency into Avril, Bob and Susan. Even Don was beginning to share their concern. Jim's appeals to Malcolm at progress meetings to 'do something about marketing before it's too late' met with underwhelming enthusiasm from many in the team. Jim no longer attended the weekly progress meetings, preferring instead to work on his own materials. Over a pint in the pub after work, Joan would refer to Malcolm as 'a wasted space' while Jim would mutter that 'he couldn't manage his way out of a paper bag'.

Joan tried valiantly to twist Malcolm's arm, appealing at once to his vanity and to his common sense: 'We're going to have to exercise much stricter control over authoring and production if we're ever going to generate enough income before the sponsorship runs out. Why don't you give me the authority to manage authoring and production while you concentrate on marketing and sales?' Malcolm's response was to hedge. 'Well, we're not sure which way the College is going to jump; whether they see us as a self-financing unit producing and selling course materials or whether we're regarded as an integral part of the academic structure that recruits bona fide students'.

Until this moment, Joan, Jim and Don had understood the project definitely to be in the business of producing and selling course materials. Now there was some doubt. Malcolm revealed discussions were underway with the College's Academic Board to redesignate the project as providing bona fide courses partly through open learning materials. As if to underline the difference, whereas team members had become used to thinking and talking in terms of 'customers', a flurry of memoranda from the College's Finance and Administration department told Malcolm that the project team were not 'publishers' and under no circumstances was the team to describe purchasers of its materials as 'customers'; they were 'students'. The project's cash-flow was also quite unlike the normal year-on-year balancing of books within the College; the prospect of a deficit in the third year of a project (not unusual in businesses) could not be countenanced by the Finance people; was it not set down in tablets of stone that no department can end the financial year with a deficit?

All this news was joy to the ears of Avril, Bob and Susan. They had been increasingly resistant to Joan's attempts to impose some discipline on their way of working and to extract some adherence to deadlines. Their argument was that they were academics and as such could not be answerable to an 'administrator'. As subject experts, they considered that they alone had to decide whether or not they had injected sufficient thought, care and attention to the material. They also argued that as academics they had no role to play in the sales and marketing of the materials. It became clear to Joan and Jim that they were operating a kind of 'work to contract'.

Avril and co. were, in effect, thumbing their nose at Joan, Jim and Don. The difference in values between the two groups could not have been more marked. Whatever the shared values that had brought them altogether in one project, they had merely overlain a fundamental gulf in beliefs about personal responsibilities and employer–employee relations, and in perceptions of the project's purpose.

Malcolm chose to sit on the fence in the growing dispute between the two factions. Instead, he spoke about the project's 'glorious future' and acted as if there was nothing to worry about. While Joan, Jim and Don soldiered on with their own work and started to publish materials, progress from Avril, Bob and Susan was still minimal. Marketing and sales activity was non-existent. By spring of 1986, the project was out of control. The sponsors notified both Malcolm and the College of their concern about an impending financial crisis. Systems consultants were called in to sort out the mess. After an initial reading of the situation, they advised against a 'quick fix' and embarked upon a soft systems study.

Step 2: Analysis

A rich picture

Figure 12.1 captures the unstructured problem situation in rich picture form.

Issues and primary task areas

The primary task of the Unit was to publish training materials. The following appear to be issues and poorly fulfilled primary task areas in the problem situation:

Issues	*Primary task areas*
staff competencies	production control
motivations, attitudes	marketing
and values	financial control

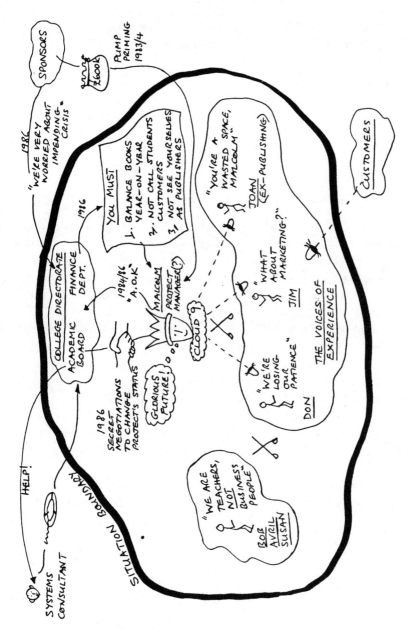

Fig. 12.1 A rich picture of the problem situation at Northbrook College Training Materials Unit.

nature of project	strategic planning
• teaching?	business plan
• business?	
• publishing?	
roles	
leadership	
lack of clear objectives	
poor communications	

Step 3: Relevant systems and root definitions

A number of the issues listed above are related. For example, differences clearly exist in motivations, attitudes, values and staff competencies. Lack of leadership and poor communications allow these differences to fester. A 'quick fix' would be to replace Malcolm with a more competent manager but this would only deal with a prominent symptom. What is clearly lacking in the project is any adequate sense of team-work. The project team is a team in name only. So, one relevant system would be 'a system for team building'. Addressing the particular issue of staff competencies, another relevant system might be 'a system to harmonise staff skills'. On the issue of poor communications, a further system could be 'a system to develop team member assertiveness'. Different understandings of what the project is about mean that different factions are motivated differently and work differently. A 'system to rationalise the project's identity' would be relevant.

Among primary task areas, lack of a business plan and a marketing plan appear significant as does 'open loop' production control. Thus, the list of relevant systems is now:

issue-based
RS1: a system for team building.
RS2: a system to harmonise staff skills.
RS3: a system to develop team member assertiveness.
RS4: a system to rationalise the project's identity.

Primary task based
RS5: a system for business planning.
RS6: a system for marketing.
RS7: a production management system.

Owing to lack of space, we are only going to develop root definitions for two of the candidate systems – RS1 and RS4.

RD1 (first pass): a system owned by the Project Manager for improving the work of the project team whereby team members develop an agreed set of team values, roles, goals and ways of working that make the best use of their individual competencies.

Inspection test – RD1
- What does 'improving the work of' mean? Tighten up wording.
- 'Ways of working' is a bit woolly.
- How will members know that agreement has been reached? Will it have to be written down for future reference/development?
- Is 'best use of' too qualitative?
- Will individual competencies need to be specified and agreed?
- Can the Project Manager's leadership be assumed?

RD1 (first iteration): a system owned and operated by the Project Manager for improving the work of both the individual members and the project team as a whole (as measured by performance criteria) whereby team members develop and record an agreed set of team values, roles, goals and work methods and procedures that make the most effective and efficient use of their individual competencies and needs.

CATWOE test – RD1
Customers: Project Manager and other team members are implicit beneficiaries; Northbrook College, the sponsors and paying customers would also stand to benefit indirectly from improved teamwork.
Actors: Project Manager and team members (explicit).
Transformation: uncoordinated and misdirected individuals converted into a coherent, unified working team.
Weltanschauung: it is implicit that team work is desirable and that team building is a legitimate process.
Owners: the Project Manager explicitly owns and operates the system, even though by definition the other team members own a stake in it; it is assumed that without a formal leader the team cannot function.
Environment: constraints include the College's Directorate, trade unions (e.g. NATFHE), the sponsors, customer demands, and time.

RD4 (first pass): a system owned by Northbrook College and operated by the Project Manager and project team for making explicit differing views of the project's identity, evaluating their relative merits, proposing viable identities, and selecting an identity acceptable to all those with a legitimate interest.

Inspection test – RD4
- Northbrook College is too vague; perhaps the Directorate or even the Director.
- Evaluating merits according to what and whose criteria?
- Viable according to what or whose criteria?
- Do the criteria for 'acceptable' need specifying?
- Define 'all those with a legitimate interest'?

RD4 (first iteration): a system owned by Northbrook College's Directorate and operated by the Project Manager and project team for making explicit differing views of the project's identity, evaluating their relative merits according to objective criteria agreed with the team, proposing identities that meet those objective criteria, and selecting an identity acceptable to all those with a legitimate interest at project level, at Directorate level and at sponsor level.

CATWOE test – RD4
Customers: Project Manager, team members, the Directorate and the sponsors are explicit beneficiaries.
Actors: the Project Manager operates the system with the active participation of team members; others with a legitimate interest at Directorate level also have a role to play.
Transformation: the essential process is the forging of a single agreed identity for the project from a variety of differing and possibly conflicting perceptions of the project's identity.
Weltanschauung: the assumption is that all actors will value the benefits of a single agreed identity for the project in preference to a variety of different and possibly conflicting identities.
Owners: Northbrook College's Directorate are the explicit owners as they legitimate the existence of and have ultimate authority over the project; they seek to ensure that the project's identity is compatible with the College's identity, constitution and goals.
Environment: constraints include the sponsors, paying customers who respond to the project's image which is influenced by its identity, and time.

Step 4: Conceptual modelling

In view of space limitations, we have only carried forward RD1. In order for RD1 to function, such a system would have to include the following main processes or verbs:

identify:	individual skills and needs
draw up:	a list of team objectives
compare:	team needs with individual inputs
specify:	key team roles and tasks
allocate:	individuals to roles and task responsibilities
agree:	set of team values, goals, work methods and procedures
plan:	a schedule of team work
carry out:	the teamwork schedule
monitor:	progress towards team objectives.

Figure 12.2 shows a conceptual model of RD1.

Inspection test

(1) Do the model and RD1 suggest a continuous and relevant mission? (A: Yes, team building and team work are obviously needed as a permanent feature of such a project.)
(2) Can performance be measured? (A: Yes, establishment of criteria, and monitoring and control procedures are included.)
(3) Is a decision-making activity present? (A: Yes.)
(4) Do any of the main activities comprise sub-systems of back-up activities? (A: Yes.)
(5) Do the system components interact? (A: Yes.)
(6) Does the system interact with an environment? (A: Yes, the College Directorate, NATFHE, the sponsors, paying customers.)
(7) Does the system have a boundary? (A: Yes, the wider system is the College Directorate.)
(8) Can the wider system provide resources? (A: Yes.)
(9) Can the system be sustained? (A: Yes, the Project Manager has formal authority to sustain the system if he or she desires.)

Expanded conceptual model

Each main activity was examined in turn to identify back-up activities that logically would be required for the main activities to function. These were listed out prior to drawing an expanded conceptual model

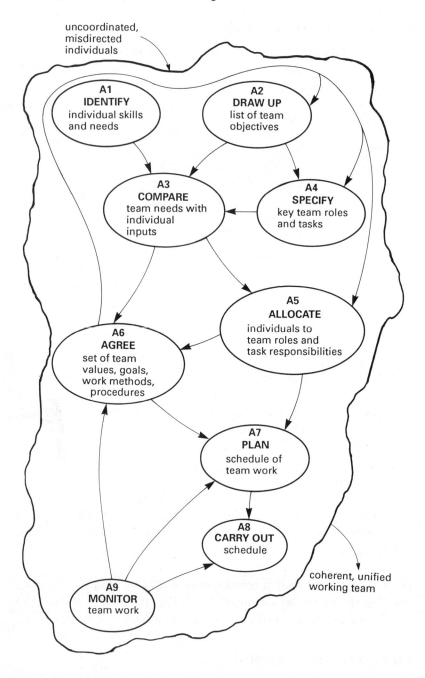

Fig. 12.2 Conceptual model of the team building system for Northbrook College's Training Materials Unit (second iteration).

as in Fig. 12.3. The expanded model was tested with the nine-point schedule and it was decided that no further expansion was needed.

Step 5: Comparison

The table in Fig. 12.4 compares the expanded conceptual model with the real-world problem situation. From the comparison table, an agenda of issues was drawn up as follows:

(1) Reconciling individual skills and needs with what would be needed to meet team objectives.
(2) Matching what team members can do and want to do with team roles and task responsibilities.
(3) Agreeing a set of team values, goals, work methods and procedures among a set of single-minded individuals.
(4) Planning a schedule of team work that everyone will adhere to for the good of the team.

Step 6: Debate

A sanitised version of the rich picture (Fig. 12.1) was not presented to Malcolm and the team as the situation was judged to be too sensitive. The agenda was first discussed with Malcolm on his own to gauge his reaction and to see whether in principle the agenda had his support. Testing the water like this is an important part of confidence building. It demonstrated that the analyst was trying to be constructively critical and allowed Malcom time to prepare for the forthcoming wider debate with other actors. The second stage involved a joint meeting between the analyst, Malcolm and representatives of the College directorate to apprise them of the agenda and allow a free discussion. Again, the aim was to test support for the 'whats' of the agenda. The final stage was a debate involving the analyst, Malcolm and the staff of the Unit using the conceptual models to steer the debate. The following changes were agreed:

(1) Using the conceptual models as a guide, Malcolm would set up formal mechanisms to reconcile individual vs. team conflicts, to set team objectives, and as far as possible to match team members' attributes with the requirements of meeting team objectives.
(2) Although it was recognised that team values were needed, it was felt that this would emerge from the formal mechanisms rather than requiring a special formal process of its own.

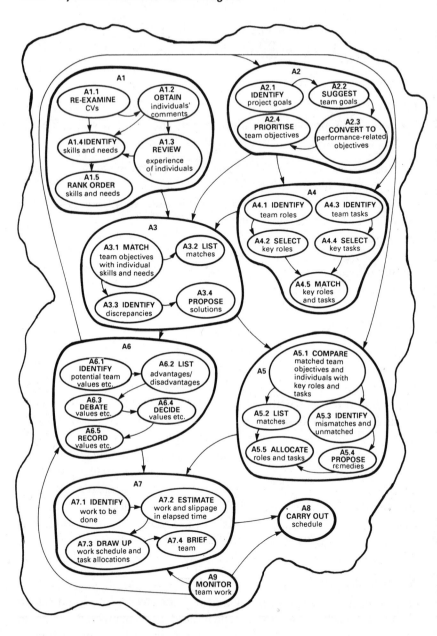

Fig. 12.3 Expanded conceptual mode of the team building system for North-brook's Training Materials Unit.

Conceptual	Model	Real-world	Observation who? how? when? etc.	Add to agenda? Y/N
Main activity	Back-up activity	Activity present		
A1	A1.1	No	Only done on hiring	Y
	A1.2	No?	No purposeful interviews by Malcolm	Y
	A1.3	No	Progress meetings not suitable for this	Y
	A1.4	No?	Not in meaningful sense	Y
	A1.5	No	Amorphous at present	Y
A2	A2.1	No	Inside Malcolm's head?	Y
	A2.2	No	Do not exist	Y
	A2.3	No	" " "	Y
	A2.4	No	" " "	Y
A3	A3.1	No	Skill matching rests on original hiring process	Y
	A3.2	No	" " " "	Y
	A3.3	No	" " " "	Y
	A3.4	No	" " " "	Y
A4	A4.1	No	Only by guesswork and emergence	Y
	A4.2	No	Only be gravitation	Y
	A4.3	No	Guesswork and	Y
	A4.4	No	assumptions	Y
			" " " "	
	A4.5	No	Malcolm's whim	Y
A5	A5.1	etc.		

Fig. 12.4 Table comparing expanded conceptual model of the team building system with the real-world situation at Northbrook's Training Materials Unit.

(3) Malcolm agreed that Joan should takeover the scheduling and with full authority for progress chasing; Avril, Bob and Susan agreed to this, although somewhat reluctantly.

Step 7: Action

One of the outcomes of the debate was a general recognition that all team members needed to adopt a much more disciplined way of doing things. A series of team meetings were set up by Malcolm each with a clearly defined agenda dealing with only one topic at a time. Team members were told that each meeting had to secure a list of agreements. Proper minutes were kept with an action column. A 'team objectives' statement was agreed and at the beginning of each meeting members were reminded of its message. When new jobs or problems arose, Malcolm arranged 'team briefing' sessions to discuss and agree a plan of action. Each member was given agreed performance targets and progress was monitored both by Joan's Gantt chart and at Team Performance meetings.

The effects of team building eventually became apparent in improved output and quality of work and a better 'atmosphere' among the team. Malcolm concentrated his efforts on managing the project and temporarily forsaking his fantasies about the future. However, it remained a tense and uneasy time for those involved and we have not pursued the outcome owing to space limitations.

12.3 Westgate Polytechnic's Environmental Health Department – a study of staff discontent

Step 1: Data collection

Environmental health per se is concerned with how people's health is influenced by their surroundings. Environmental health specialists are concerned with monitoring and controlling the causes of potential ill health in the environment. Typical subject areas are food hygiene in shops and restaurants; the health and safety of working conditions in offices, shops and warehouses; pollution of the public environment by noise; the quality of rented housing accommodation; and pest control.

All these areas are the subject of statutory legislation, for example the Public Health Acts, the Health & Safety at Work Act, the Food

Hygiene Regulations, the Housing Act, and the Control of Pollution Act. The enforcement of this legislation is delegated mainly to Environmental Health Departments of local authorities who employ Environmental Health Officers (EHOs) to inspect premises and activities, issue enforcement notices, monitor compliance and prosecute occupiers if necessary. In earlier times, EHOs did not exist; there were Public Health Inspectors whose scope of activities was much narrower and covered such topics as the adequacy of washing facilities and toilets in offices, shops and restaurants. Nevertheless, the longstanding traditions of Public Health Departments, namely emphasis on 'policing', are part of the culture of older Environmental Health Officers.

The education of EHOs traditionally comprised a 'thin sandwich' Diploma course studied at a Polytechnic or College of Technology. Typically, six months would be spent at college followed by six months back at the student's sponsoring EH Department of a local authority. Such a course lasted three years. However, their professional body, the Institution of Environmental Health Officers, began to aim for an all-graduate profession and by the early 1980s the pressure on colleges to offer a degree course was growing.

The Diploma course typically addressed what colleges saw as the practical needs of practising EHOs. Lecturers were drawn largely from among former EHOs whose wealth of experience was highly regarded. However, the mix of academic staff in Environmental Health Departments in colleges tended also to reflect the range of legislation and subjects taught. It was common to find former EHOs and other lecturers having a variety of technical and scientific backgrounds all sharing the Department's staff room. Westgate Polytechnic's Environmental Health Department (EHD) fitted such a mould.

The pressure to switch from the Diploma to a BSc course was only one of several stimuli for change that hit Westgate's EHD in the mid-1980s. For the previous 20 years, the Department had been headed by Jock a former EHO. Over that period, the EHO contingent of the Department reigned supreme but in the last few years, especially with the advent of a lot of new legislation concerning hazardous substances, the proportion of scientists had increased. When Jock retired in 1986, the scientists numbered three, and the EHOs and assorted others, ten.

The EHOs assumed that the progression was assured and that one of them would be appointed as Jock's successor. They felt that Jock had given them every indication that this would happen; but Jock was nothing if not canny. He knew that for the Department to survive the multiplicity of major changes ahead it would have to have a very different kind of head than he had been. Jock ensured that Mike, one of the scientists, got the job.

Mike held a PhD as did one of the other two scientists. As posts became vacant or new posts were created, he began to fill them with PhDs or MScs, thus endorsing his view that the department's credibility depended on having a more academic and scientific approach to the subject. He was mindful of increasing competition in the higher education sector and the implications of the proposed Education Reform Act.

The non-scientist faction was very resentful of these developments but did not seem to know how to deal with them. When Mike stated at staff meetings what he intended to do, most of the detractors either said nothing or indicated their support. However, discontent was expressed openly in the Department staff room but only when Mike was absent. The staff room in effect became the 'catharsis room' where all their anger and frustration was poured out. There were many willing ears.

The scientist faction remained aloof from the back-biting, even when on the receiving end of caustic comments. The ex-EHOs and other technologists felt that their traditional vocational approach was still the correct one and was valued among local authority Chief EHOs. 'An ounce of practical experience is worth a ton of theory', one of the disaffected was heard to proclaim. Nevertheless, they were getting increasingly uneasy at what they saw as the ambivalent attitudes of such employers and the EHOs' professional body. Traditional EHO roles were being questioned. Should EHOs wield a big stick or should they encourage self-regulation? Should they be 'policemen' or advisers? Should they continue to deal with traditional tactical matters or should they focus on strategic management of environmental health?

Mike, it has to be said, was a forceful chap but he had no management training or experience. He tended to treat people as if they were simple, predictable beings whose compliance with his wishes would be straightforward and not involve adverse emotions. When, after his first two terms, it became clear that there was an increasingly awful atmosphere in the staff room and staff meetings were sullen, uncreative affairs, Mike was at a loss to know what to do. A soft systems analyst was asked to study the problem.

Step 2: Analysis

A rich picture

Figure 12.5 is a rich picture of the problem situation.

Issues and primary task areas

Clearly, the problem situation resides in an environment of change and uncertainty. The perceived ambivalence of the Institution and Chief EHOs towards the role of EHOs fuelled uncertainty among the staff about what their approach to teaching should be. Changes in technical legislation as well as proposed education reform added to the ambiguity of their position.

'Them-and-us' factionalism had developed almost as a knee-jerk response to a rapid, unexpected change, namely the ascendance of the scientists. Adherence to micro-cultural or partisan values created an illusion in some that changes could be resisted or even reversed.

Mike had a steamroller tell/sell approach to 'man management' that made staff feel impotent to influence decisions. Apart from Mike's edicts, the management system of the Department was unclear to the staff.

In summary, the following issues are identifiable. The analyst did not consider that there were any primary task areas of importance in this case:

> *issues*
> coping with uncertainty
> departmental management
> partisan loyalties
> coping with change
> approach to teaching
> decision making

Step 3: Relevant systems and root definitions

The following are relevant systems:

RS1: a system for collectively coping with uncertainty about topics of major importance to the Department.

RS2: a system for reconciling partisan loyalties among staff.

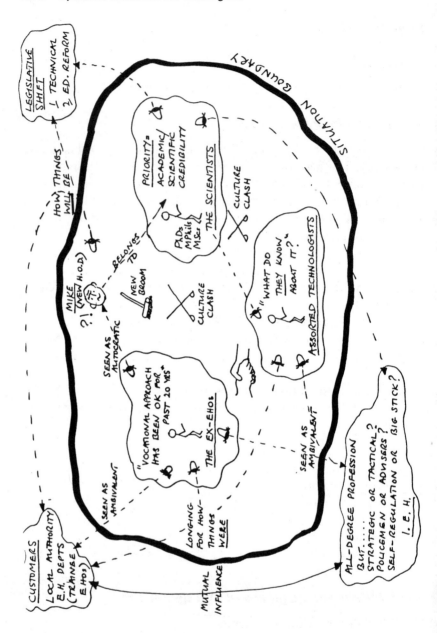

Fig. 12.5 A rich picture of the problem situation at Westgate Polytechnic's Environmental Health Department.

RS3: a system for coping with major changes that might affect the Department.
RS4: a system for reconciling vocational and academic approaches to teaching.
RS5: a system for consultative decision making.
RS6: a system for managing the Department.

In view of space limitations, we are only going to develop RS1 whose root definition may be expressed as:

RD1 (first pass): a system owned by the Head of Department and operated by all the teaching staff for identifying uncertainty about major topics of importance to the Department, defining the extent of uncertainty, assessing its potential implications, devising ways of reducing the uncertainty, and applying means for coping with the residue of uncertainty.

Inspection test
• Could there be more than one uncertainty about a topic?
• What criteria will be used to decide what are major topics?
• Who decides what is important?
• What measures of uncertainty are there?

RD1 (first iteration): a system owned by the Head of Department and operated collectively by all the teaching staff for identifying uncertainties about topics agreed collectively to be of major importance to the Department, defining the extent of uncertainties in both quantitative and qualitative terms, assessing their potential implications in both quantitative and qualitative terms, devising ways of reducing the uncertainties, and applying means for coping with the residue of uncertainties.

CATWOE test
Customers: the Department as a whole; indirectly the Department's students, their employers, and Westgate Polytechnic.
Actors: explicitly all the teaching staff (including Head of Department).
Transformation: the essential process is one of modifying perceptions of uncertainty, specifically to reduce uncertainties and cope with residues.

Weltanschauung: the view that unstructured, unmanaged uncertainty is undesirable as it leads to frustration and anxiety.

Owner: the Head of Department explicitly owns the system, i.e. has power to end its existence.

Environment: there are numerous components that affect and constrain the system, many of which are sources of uncertainty for the Department, e.g. Westgate Polytechnic authorities, Government intentions towards higher education, technical legislation, local authority Chief EHOs, economic climate, the IEH's attitudes towards professional competence, public attitudes towards environmental health.

Step 4: Conceptual models

From RD1, the following main verbs are indicated for the essential logical functioning of the system:

agree:	topics of major importance
identify:	uncertainties about topics
define:	extent of uncertainties
assess:	potential implications
devise:	uncertainty reduction mechanisms
reduce:	uncertainties
isolate:	residues of uncertainties
apply:	coping means to residues.

A conceptual model of RD1 main verbs is shown in Fig. 12.6.

Inspection test

(1) Do the model and RD1 suggest a continuous and relevant mission? (A: Yes, uncertainty is a permanent feature of the Department's existence and the way uncertainty is managed affects the Department's degree of success.)

(2) Can performance be measured? (A: Yes, since the extent of uncertainties have to be measured and their implications assessed before and after reducing and coping activities occur, performance measurement is implicit.)

(3) Is a decision-making activity present? (A: Yes, agreement implies collective decision; devising reduction mechanisms and applying coping means imply an element of decision-making.)

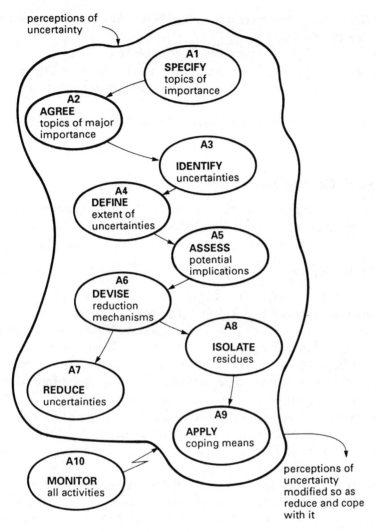

Fig. 12.6 Conceptual model of main activities in the uncertainty management system for Westgate Polytechnic's Environmental Health Department (first iteration).

(4) Do any of the main activities comprise sub-systems of back-up activities? (A: Yes.)
(5) Do the system components interact? (A: Yes.)
(6) Does the system interact with an environment? (A: Yes, see CATWOE test.)

(7) Does the system have a boundary? (A: Yes, the wider system is the Polytechnic.)
(8) Can the wider system provide resources? (A: Yes.)
(9) Can the system be sustained? (A: Yes, the Head of Department has formal authority to sustain the system if he or she desires.)

Figure 12.7 shows an expanded conceptual model that includes back-up verbs.

Step 5: Comparison

A tabulated comparison revealed matches and mismatches between conceptual model and real-world problem situation.

Step 6: Debate

Mike was presented with a sanitised rich picture and the proposed agenda of issues relating to RD1 as a prelude to a wider debate. He agreed with the essence of the analysis and the 'whats' of the agenda. It was agreed that an open debate with him and his staff was called for. The debate did not go well. There were mutterings about 'psycho mumbo-jumbo' from some quarters. It was clear that deeply entrenched value systems were operating. The analyst abandoned RD1 and switched to an agenda and conceptual model for RD2 (which we have not developed here). The debate fared better with RD2 and there was a measure of agreement that 'something' was needed to weld a unity of purpose that would subsume partisan loyalties. RD2 and its conceptual model were not unlike the 'team building' system considered in the Northbrook case and the actors agreed to consider the requirements of putting RD2 into practice. RD6, the management system, was also debated and received a favourable response.

Step 7: Implementation

The debate had highlighted just how little any of the lecturers, including Mike, knew about management. Before attempting to implement RD2 and RD6, it was agreed that all the lecturers would begin a programme of management courses covering such topics as assertiveness, project planning, stress management, negotiating and so on. This helped them view the management of the Department as a

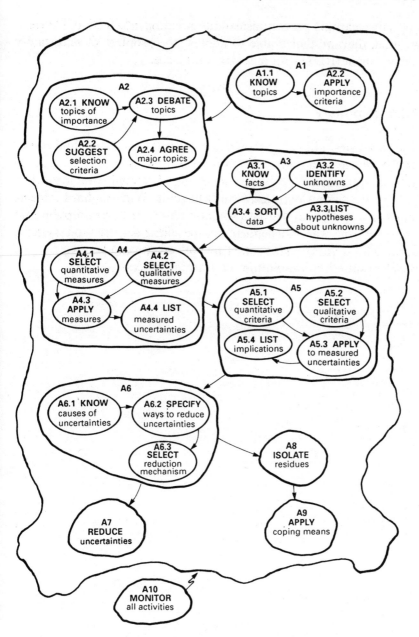

Fig. 12.7 Expanded conceptual model of the uncertainty management system for Westgate Polytechnic's Environmental Health Department.

collective learning process instead of retreating into factional laagers. In time, they were also able to address the problem of uncertainty management (RD1) with renewed confidence.

12.4 Summary

The two case studies have shown how the soft systems method can be used to navigate a path through what at first sight may seem to be almost intractable problems involving factional disagreements, personality clashes and defensive behaviour. Whereas hard systems analysis would try to deliver a solution based on an assumption that 'the problem' was independent of the client set, the soft systems approach assumes that the actors are both creators and resolvers of the problem situation. Resolution is a learning process.

Chapter 13
Analysing Systems Failures

13.1 Introduction

Chapters 7 and 8 in Part 1 introduced the concept of system failure and showed how you could use various models to examine aspects of system failure. However, as noted in Chapter 8, simply making comparisons with models in a haphazard way does not represent a systems approach. Systems failures analysis provides a disciplined way of examining failures that often reside in quite complex, messy situations. This method enables the analyst to gain understanding of situations that exhibit apparent failures and to draw lessons for preventing similar failures. The degree of understanding gained is determined largely by the number of iterations carried out which in turn depend on the analyst's resources – especially time. However, even a single pass through the method can often be quite illuminating.

13.2 Systems failures analysis

The essence of systems failures analysis (SFA) is comparison between the apparent failure situation and a range of models. Some models such as control and communication are desirable in the real world whereas others such as fault trees and cascades are undesirable. If elements of desirable models are missing or elements of undesirable models are present in the situation, it suggests possible contributions to the failure from which lessons may be learned.

Figure 13.1 depicts the stages of SFA. Your starting point is to describe the failure situation. However, prior to this you need to outline your objectives even though they may be vague. The point of this is to (a) make you think about what you are doing, and (b) make

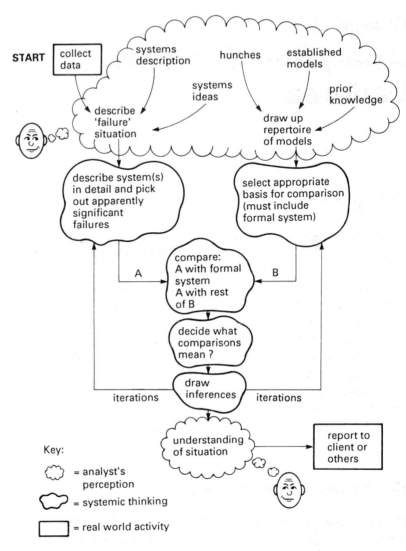

Fig. 13.1 Procedure for analysing systems failures. (Adapted from Joyce Fortune, *Studying Systems Failures*, Open University T301 Course Material 1984).

you pay attention to your resources. Once you begin to describe the failure situation, you may well find that you will have to limit the scope of your study and revise your objectives. An objectives hierarchy is a good way of focussing your plan of work.

13.3 Describing the failure situation

The situation you are studying may be quite local and self-contained but could easily be very complex, widespread and have many ramifications. Whatever the case, you will be searching for evidence that someone at least is dissatisfied with the performance of something that seems relevant to your task. Your own world-view and those of actors in the situation are obviously pertinent to the perception of failure and what is relevant.

Searching for relevant data involves obtaining relevant documents, interviewing relevant people and maybe visiting places where relevant failures are said to have occurred. Both at this stage and when reporting or during later iterations, you need to avoid using the term 'failure' or 'failures method' when communicating with relevant people. The fact that failures are apparent does not necessarily mean that everyone shares that perception and it is likely that 'failure' will be taken as a sign of criticism. Euphemisms like 'difficulties' and 'problems' may be appropriate.

Having gathered what you think is enough information, draw, say, a rich picture to cover as much of the failure situation as possible. A spray diagram can be just as useful. Some purists argue that rich pictures should only be used in soft systems analysis. We disagree. Doing this pre-analysis sketching should suggest particular areas to focus on from which you can begin to tease out specific systems to study.

Case study for this chapter

For the rest of this chapter, you will have to do most of the work. If you get stuck at any point, refer to the comparable parts of the worked case studies in Chapter 14.

In Chapter 9, we referred to the seemingly intractable problems of air travel and air traffic control. July 1988 saw European flights to the Mediterranean delayed on an unprecedented scale. Media interest was substantial and just one front-page newspaper article with the banner headline 'CAA Heads Must Roll' may be cited to convey the extent of failure (*The Observer*, Sunday 17 July 1988). The article suggested that a political row was threatening to break over the Civil Aviation Authority as government ministers blamed it for causing massive airport delays 'which could become routine for at least the next seven years'. Phrases such as 'mistaken forecasts', 'capitulation to chaos', 'the debacle', 'failed to prepare for the surge in flights', 'crisis', 'hell of a mess of things', and 'past failings' are all indicative of apparent failures.

Gather a range of data about the air traffic control situation from newspaper articles, official reports, airport visits or whatever sources you have readily accessible. Construct a rich picture and/or spray diagram of the unstructured situation. Then carry out the steps of the method as described below.

The next step in describing the situation is to see if there are any systems in it that are worth studying. Here you need to use the systems description as described in Chapter 2. Find a topic area in your rich picture or spray diagram that looks relevant to the failure situation; you may find several. An example might be air traffic control per se. Another might be flight safety. As you are using this case for study purposes rather than as a consultant working for a client, pick out a range of likely problem owners and construct a commitment statement for each.

Now test each commitment statement to see whether it is goal-orientated, whether analysis is more important than action, whether analysis is need-driven rather than curiosity-driven, and whether the problem owner's task is relevant to the failure situation. In other words, are any of the problem owners dealing with problems that are systemic or system-borne and therefore would benefit from failures analysis? Examples of how to carry out the test are given in Chapter 2.

Having found at least one key problem owner, focus on two or three areas that your key problem owner would be interested in.

Exercises

1. Suggest three of four areas that the Director of Air Traffic Services at the CAA would almost certainly be interested in.

Now focus on one of the areas and separate from it at least one system that exhibits apparent failures. For example, if you chose the operation of air traffic control centres then one system that exhibits apparent failures is the centre at West Drayton that covers most of the air traffic in and around southern England. All systems need a title that reflects ownership and purpose and so a title might be 'the CAA system for operating the air traffic control centre at West Drayton'.

From your collected data on the ATC centre at West Drayton, draw another spray diagram or rich picture of the situation. You may find the information contained in Figs 5.4 and 5.6 in Chapter 5 useful, although it is not a substitute for your task in this chapter. Then list out some likely system components and divide them with a trial boundary between 'within the system' and 'in the environment'. Repeat the process of system separation, rich picture construction, and trial boundary for the other areas of interest.

You now have several potential systems to examine but are any of them crucial to understanding why failures occur in air traffic control? Check back against the commitment statement you constructed for your key problem owner (i.e. the Director of Air Traffic Services). Then peruse your data to see if there are any apparent failures that relate to this problem owner's task and which suggest models for comparison. For example, evidence of lack of monitoring and control are key indicators of failure. Accidents and structural failures are prima facie evidence. The formal system model, control model and fault tree paradigm would, for these examples, illuminate particular aspects of the failure situation of each key system.

13.4 Detailed systems description

By now, in effect you will have completed your first quick pass through the method and know whether or not you have a key system that seems to exhibit failures that are significant to your problem owner's task. The next stage is take a key system and describe it in detail.

Components

Examine the components listed when you first separated the system. Are there any important ones missing? Do any need to be transferred from inside the system to its environment or vice-versa? Are all the components at the same level of resolution? For example, Fig. 13.2 shows one iterative adjustment and you can see that part (b) is more structured and easier to understand. Aim to keep the number of main components to less than 12.

Summarise the relationship between the components in the form of

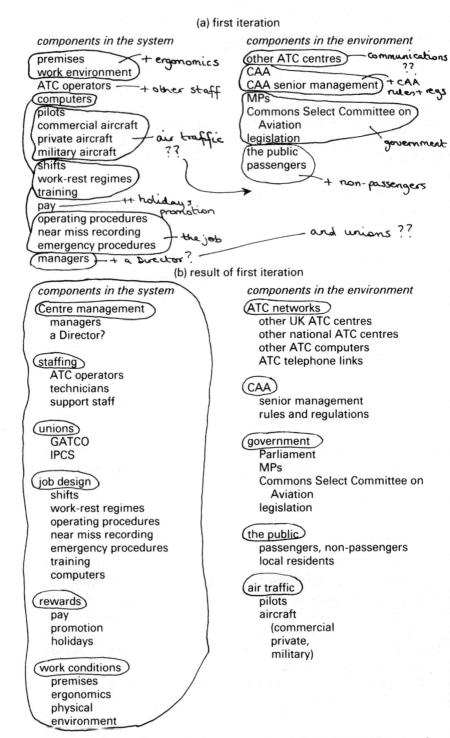

Fig. 13.2 Iterative adjustment of component resolution for the CAA system for operating the air traffic control centre at West Drayton.

a system map. Several iterations may be needed before you feel happy with it. What was once a foggy impression of the system should be starting to look clearer.

Inputs

What ingredients enter this system in order for it to function? List them out and adjust the resolution as necessary.

Outputs

What are the outputs of this system? There are obvious outputs such as the majority of aircraft reaching their destinations safely and some of them on schedule. Equally obvious is the unwanted output of flight delays.

Exercises

2. List other outputs from the West Drayton ATC Centre system.

Environment

What are the components outside the system that affect it? Government, the media, air passenger volumes and other ATC centres are obvious components. CAA senior management are also outside the West Drayton Centre system and represent its wider system.

Structure

What are the major sub-systems that make up the system? Look for decision-making, monitoring, control, and operational sub-systems and list them out. Your system map will help but if your collected data do not tell you then you may have to seek some more.

Variables

The variables are those components whose contributions to the system's properties may vary. For example, in the West Drayton ATC Centre system, job design will vary because sub-components such as training, work-rest regimes and operating procedures may change. List out the variables and adjust the resolution as appropriate.

Relationship between variables

Here, all that is needed is an influence diagram showing how variables affect each other and an indication of where the boundary lies.

Apparently significant failures

There are a number of ways to tackle this step systematically. One is to consider failures that occurred during different phases of a system's life. Another is to categorise apparent failures according to the functions or sub-systems they reside in. The case studies in Chapter 14 show both kinds of categorisation. Usually you have identified quite a number of apparently significant failures and putting them in categories makes them more manageable.

In the West Drayton ATC Centre system, for example, you may find apparently significant failures that lend themselves to categories such as resource provision and implementation, monitoring, equipment failures, and so on.

Appropriate models

With organisations the formal system model (FSM) is a necessary model to select for comparison (see Fig. 7.1 in Chapter 7). In addition, select appropriate models for each failure aspect of each failure category identified. For example, a particular aspect of equipment failure may suggest the fault tree and/or the cascade model. Particular monitoring failures may suggest the control model.

13.5 Comparison

Select a number of apparently significant failures and corresponding models for comparison. Which ones and how many you select are a matter of judgement and the time you have available. Comparison takes the form of a template exercise. In other words, take the failure situation or particular failure and superimpose it on the appropriate model diagram. Chapters 7 and 8 depict a number of important models and Chapter 14 shows examples of template comparisons.

13.6 What do comparisons mean?

In this stage of analysis, you will be interpreting the comparisons from the previous stage. Always interpret the FSM comparison first. Typically, deficiencies in sub-system structure or processes are highlighted, e.g. monitoring sub-system absent or ineffective. Monitoring and control defects highlighted by the FSM comparison then need to be checked by comparison with the control model.

A simple 2 × 2 table is useful for interpretation. Bear in mind that some models are desirable and others undesirable and so in principle there are four possible combinations of whether the model is desirable or undesirable and whether the comparison reveals many or few matches. Combinations in quadrants 1 and 4 represent an 'acceptable' situation:

	Model Desirable	Undesirable
No/few discrepancies	1	2
Many discrepancies	3	4

Exercises

3. In which of the four quadrants numbered in the above 2 × 2 table would you place the following comparisons: (a) control model shows no performance monitoring, (b) control model shows maintenance sub-system not to be functioning, (c) FSM shows executive committee did not meet for many years, (d) fault tree model shows hierarchy of events leading to component failure.

Sometimes it is not possible to allocate the interpretation to a particular quadrant. This is often the case with communication failures where on the one hand there is evidence that a message was received and apparently understood and on the other hand the message was not acted upon as expected. In such cases, your mark has to float on the boundary between quadrants 1 and 3 until you are able to clarify what took place. A communication failure of this kind (i.e. behaviour of receiver inconsistent with having received and understood message) suggests one or more human factors models to consider in further iterations.

13.7 Learning lessons

This stage allows you to summarise what you have learned so far from the failures study. For example, the understanding gained about the behaviour of particular systems may enable you to suggest preventive action or treatments. The understanding may point to the need for hard systems analysis and/or a soft systems study.

You then have to decide whether you need to carry the failures investigation further and whether you have the resources, especially time, to do so. If further examination is deemed necessary, run through the method again (internal iteration) on your first key system and modify your description and analysis in the light of understanding gained in the first pass.

Further understanding may be gained by going back to the areas and key systems that you identified during systems description and follow the method through for however many (external) iterations you require.

13.8 Summary

Systems failures analysis (SFA) enables you to gain understanding about complex situations that exhibit apparent failures. It provides a disciplined approach to making comparisons between aspects of a key system in which a particular failure appears to reside and the repertoire of available models. Chapter 14 provides two worked case studies of SFA.

13.9 Suggested answers to exercises

1. The CAA Director of Air Traffic Services would be interested in:

    A: maintaining flight safety
    B: the operation of air traffic control centres

C: the use of computers in air traffic control

D: industrial relations in air traffic control.

2. Other outputs are: raised public anxiety about air safety, public anger at delays, anger among politicians, industrial relations problems, possible increased stress on controllers, computer breakdowns.

3. Comparisons (a), (b) and (c) are all discrepancies with desirable models and so your tick, cross or other suitable mark would go in quadrant 3. Comparison (d) is a direct match with an undesirable model and so your mark would be in quadrant 2.

Chapter 14
Systems Failures Case Studies

14.1 Introduction

Both the case studies in this chapter have something to do with the construction and building industries. This selection is coincidental and in no way is intended to suggest that these industries are failure ridden or suffer more failures than any other industry. There is certainly evidence to suggest that these industries have got a considerable safety problem (as in the Littlebrook D case) but this is a reflection of general weaknesses of safety management throughout employment as much as it is of any special circumstances of construction work. The lessons to be learned from these cases should be of interest to all industries and all managers. Both cases actually occurred. In the Daleside case, names have been altered to avoid offence to those involved in what was a highly controversial case. Littlebrook D, however, was the subject of an official enquiry and court cases and is a matter of public record.

14.2 Daleside Development Group – a study of failure of system-built housing projects

During the 1950s and 1960s, demand for public housing (council housing) increased dramatically. In order to reduce the time and costs of construction, 'systems building' methods were introduced by many local authorities. These methods are perhaps better described as 'kit building', for the so-called systems were little more than buildings assembled from prefabricated components. In the short-term, systems building methods achieved the aims of cutting costs and reducing the time before dwellings could be occupied. However, many serious counter-intuitive outcomes arose between the late 1960s and the mid-

1980s to the extent that system-built public housing was widely condemned as a failure.

A large number of different systems building methods using steel frame construction and prefabricated reinforced concrete were used in the 1960s. The number of houses involved were at least 170 000 plus hundreds of tower blocks, i.e. those with at least 10 storeys. By the mid-1980s, such dwellings had deteriorated to such an extent that many were considered irretrievable. Evidence suggested that many tenants had suffered ill health and general loss of amenity. Pressure from tenants' action groups which had been building up over a 15-year period had, by the mid-1980s, forced local authorities to address the problem. However, by then the scale of the failure and the costs of remedying it were enormous. Local authorities were caught between two costly options: remedial treatment or destruction and replacement. Some estimates put the cost of dealing with the tower blocks alone at between 3 and 5 billion pounds. In 1985, Manchester City Council hosted the Hulme Conference which was intended to bring together the main interest groups and outline an action plan. Following the Hulme Conference, over 100 local authorities formed the National Systems Built and Tower Block Housing Project in order to offer a more consistent approach to the failure. Many authorities have since decided to demolish tower blocks as the cheaper of very expensive options.

The analyst's client was a local authority Director of Housing. You need to appreciate that the analyst's own world-view will affect his or her perception of the situation. The analyst who actually carried out this study is a former local government officer and his view of the housing situation is affected by that experience. Whereas housing problems clearly involve and affect large numbers of people in diverse groups, the analyst took the view that major day-to-day problems of exercising responsible action in public housing rest with local authority Housing Departments. A system map of such a department is given in Chapter 2, Fig. 2.3. Figure 14.1 shows the organisation of a typical local authority housing department.

This case study concerns system building methods used in the 1960s. According to Marsh (1985), the methods as understood then were quite different from modern system building methods. Foster (1983) describes those earlier system methods as 'closed' in view of incompatibility between component designs across different methods. Current industrialised building methods use components that are

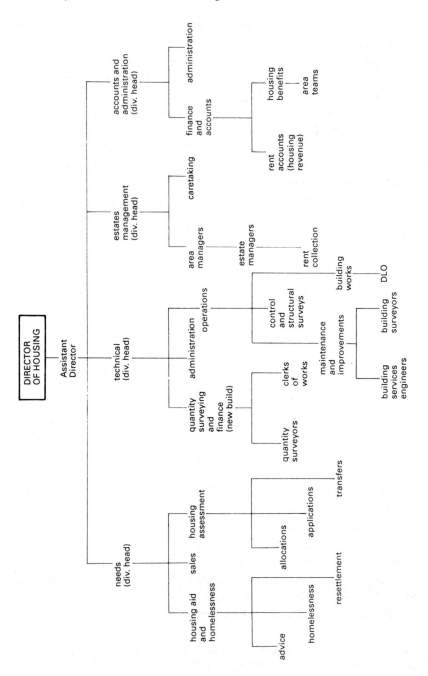

Fig. 14.1 Organisation chart of a typical local authority housing department.

interchangeable and are thus 'open'. The case study findings relating to poor engineering reliability of components refer to closed methods of the 1960s and should not necessarily be extended to system building methods in general.

14.3 Case study objectives

The hierarchy in Fig. 14.2 summarises the initial objectives of the study. However, try to avoid setting initial objectives in metaphorical concrete. Be prepared to modify objectives after your first pass or first iteration once you have had time to reflect on what you have learned.

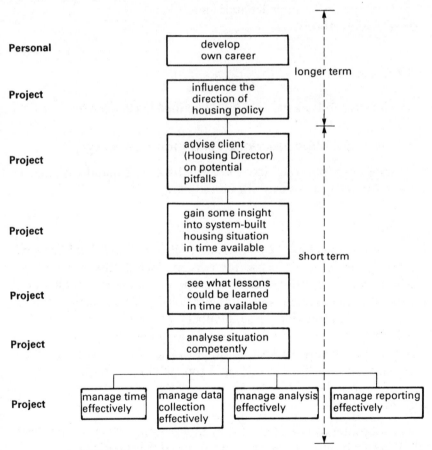

Fig. 14.2 Objectives hierarchy for examination of system-built housing situation.

Typically, a first pass through the method highlights objectives that are far too ambitious for the time and other resources at your disposal.

14.4 Data collection

The introduction above summarises the data that were gleaned from a variety of sources. Data collection methods included:

- personal observation of system built dwellings
- personal interviews of relevant people in the situation
- literature search (relevant reports and documents, reference books, press cuttings etc.)
- correspondence and telephone enquiries
- television reports.

14.5 Analysis

The analysis follows the sequence described in Chapter 13.

Stage 1: Systems description and model selection

Systems description as described in Chapters 2 and 13 is used to assemble key systems to work on as follows.

Awareness

As the introduction above indicates, all is not well with housing. Problem areas giving cause for concern include the durability of council housing, and system-built properties in particular. However, there are considerable pressures on housing departments of local authorities to provide large numbers of dwellings cheaply. For example, council waiting lists remain long and obligations exist to house the growing number of homeless people. At the same time, housing subsidies from central government have been cut.

Commitments

The following commitment statement was summarised for the client, a Director of Housing, a problem owner in the situation:

'I am responsible for assessing the needs of people who want housing from public resources. My customers include a wide range of people such as the homeless and many people on housing benefit. Once needs are assessed, households are allocated accommodation as soon as available. Of course, it does not end there because flats and houses need maintaining and repairing – tenants soon let us know the problems. With increasing demand, the housing stock has to be replenished as properties are sold off to sitting tenants or get beyond economic repair. So, I have to brief the council's Architects Department as to the functional requirements of "new build" housing, the design of which is a joint responsibility more or less. I need to understand the situation to ensure that the council effectively discharges its statutory obligations in all these respects. I am aiming for success not failure.'

Testing

- Is the client's purpose relevant to the failure situation? (Y)
- Is the commitment (of client and analyst) goal-oriented? (Y)
- Is analysis more important than action at present? (Y)
- Is analysis needed prior to action? (Y)
- Is analysis need-driven rather than just to satisfy curiosity? (Y)

The test results indicate that systems analysis is warranted.

Separation of systems

The task of a Housing Director appears to be an ideal case systemically. As a problem-owner, a Housing Director might be interested in the following potentially fruitful areas:

A: new build (including briefing, design and construction)
B: maintenance and improvement (e.g. tower blocks, refurbishment with in situ tenants)
C: housing the homeless.

A focussing topic for Area A is system building and the particular system separated for this study was entitled the Daleside Development Group Project System for 'System Building' in which the client had an interest. The Daleside Development Group (DDG) was a consortium formed in the early 1960s by four city councils in the north of England. Some 4000 DDG dwellings were system-built before the Group disbanded in 1968. The chairmen of the four councils' housing committees formed the DDG senior management but from 1968 to 1982 they did not meet on DDG business. Over that 14-year period, water seepage, dampness, mould and fungus were constant problems for tenants. Local GPs stated categorically that living under these conditions had adversely affected the physical and mental health of many tenants. Tenants had been unable to get DDG to remedy the problem and some tenants had been driven to attempt suicide. By the early 1980s, DDG dwellings which had an expected useful life of 60 years were in an advanced state of decay. In one block of flats, a three-ton parapet slab became loose and urgent action had to be taken to restrain it, so it would not fall through the roofs of adjacent dwellings. The national media were now focussing public attention on the DDG scandal.

Apart from the DDG senior management, a key figure was their designer Richard Stevenson who was responsible for both design and selecting the building contractors. It appears that DDG management simply left the designer to choose a system-building method from the 90 or so available. Swallow Construction put in a very low bid and were accepted, although Stevenson claimed that before acceptance he asked them whether they were happy with meeting the specification and whether they were sure they put everything in. He later blamed Swallow for underestimating the difficulties involved in building to his complex design. He also blamed the factories where the prefabricated panels were made as well as poor workmanship on site. Although a spray diagram could have been used, the analyst chose to summarise his perception of the DDG situation in the form of a rich picture (Fig. 14.3).

Likely components in the DDG Project System (system A) are:

Within the system	In the environment
four chairmen of housing committees	tenants
R. Stevenson, designer	Swallow Construction
briefing for designer	factory making pre-cast components
design	factory quality control
tendering	GPs
construction	Shelter
workmanship on site	NBA
repair/maintenance	tenant groups
checks on materials	DoE
fire precautions	government finance
	press/media reports

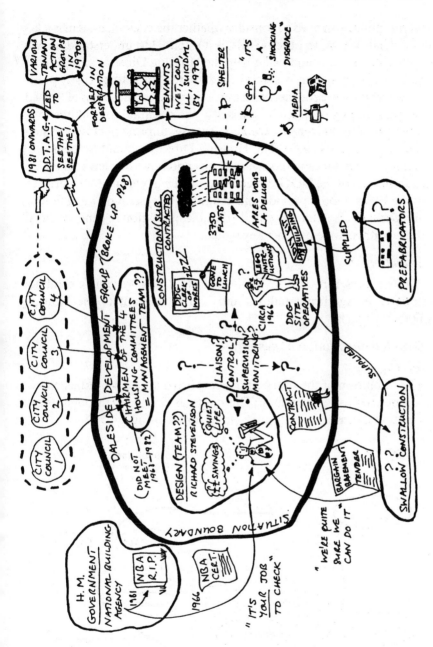

Fig. 14.3 A rich picture of the Daleside Development Group system for 'system built' housing (second iteration).

At this point, you need to consider whether the relevant system is a key or crucial system. In other words, is it essential to understanding why such failures occur and will it help the Housing Director carry out his task? As far as system A is concerned, there appear to be costly public housing failures associated with the design and construction of 'closed' system building projects. There is also the potential for failures to be repeated if it becomes expedient to use unproven closed system building methods again. The Housing Director should be helped by gaining understanding of such methods and how they have failed. An examination of the DDG Project System, being typical of such projects in the 1960s, should provide understanding of system building failures and perhaps also failures to build durable habitation in general. On all these counts, system A is a key system.

Selection

A quick pass through the method showed that system A exhibited apparently significant failures and is thus relevant to the Housing Director's task.

Detailed description of system A

(a) Components

The resolution of the trial components listed above was adjusted and summarised as a system map. This map was developed through several iterations from Fig. 14.4 to the state shown in Fig. 14.5.

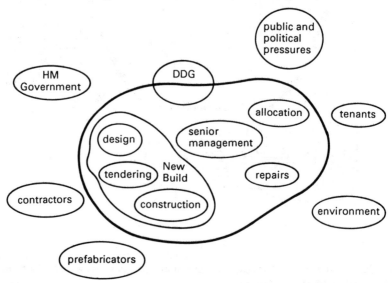

Fig. 14.4 System map of the Daleside Development Group system.

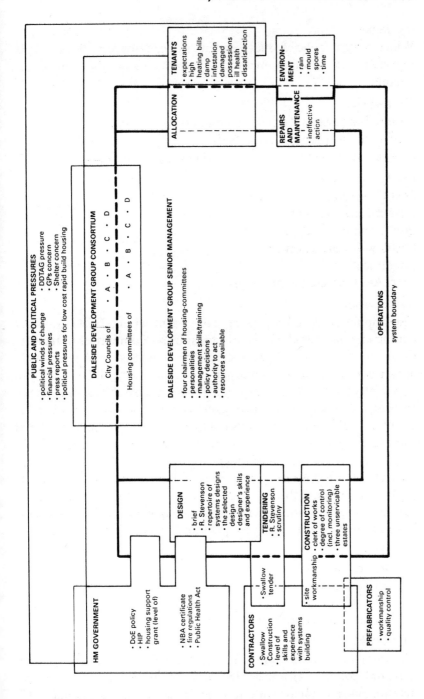

Fig. 14.5 A system map of the Daleside Development Group system from 1966 to 1982 (fourth iteration).

(b) Inputs
Wider system policies, management skills, resources, design skills, repertoire of building methods, prefabrication skills, materials, tenants.

(c) Outputs
system-built estates (largely unserviceable)
sitting tenants (largely dissatisfied)
complaints from tenants to GPs, DDTAG, the press.

(d) Environment
HM Government
physical environment
public and political pressures
Swallow Construction
prefabricators
DDTAG (the tenants action group).

(e) Structure
The appointed decision-making sub-system within DDG was that comprising the four chairmen of the Housing Committees of the City Councils in the consortium. It should be noted that unlike new build arrangements in a conventional local authority, DDG did not apparently use the Architects Departments of any of its four councils. Instead, it employed its own designer. The analyst tried to confirm the structure with the four councils concerned but confirmation had not been received by the time the case study had been completed. This highlights the importance of two practical issues in this kind of study: (i) data collection is not a process carried out only at the start of the study – it continues throughout the project as circumstances dictate, (ii) when new information is received it feeds further iterations. On the basis of the data available, the apparent DDG sub-systems were:

> design
> tendering
> construction
> allocation
> repairs and maintenance.

(f) Variables
The variables identified in the first pass are those listed below in the left-hand column. However, the level of detail was too fine and iteration allowed the resolution to be adjusted to a more manageable level as shown in the right-hand column.

First pass	Iterative adjustment
Councils in DDG } Housing Committees	DDG consortium/sponsors
chairmen of four Housing Committees DDG management and administration	DDG senior management
briefing of designer designing tendering	preconstruction
job scheduling assembly erection monitoring of work	construction
monitoring for deterioration maintenance repair	post-construction
tenants' health tenants' expectations tenants' satisfaction level of tenants' complaints	tenants

(g) Relationship between variables
In the first pass, a relationship diagram between the variables in the previous left-hand column was drawn. On iteration, the diagram was adjusted as shown in Fig. 14.6.

(h) Apparently significant failures
Four failure categories over time are discernible:

Category 1: pre-construction
The evidence suggests that the four chairmen gave the designer too much latitude in his work. There was no checking of the NBA certificate to ensure that it was appropriate to the task. Scrutiny of tenders was lax and the difficulty of the design was not fully conveyed to the builders.

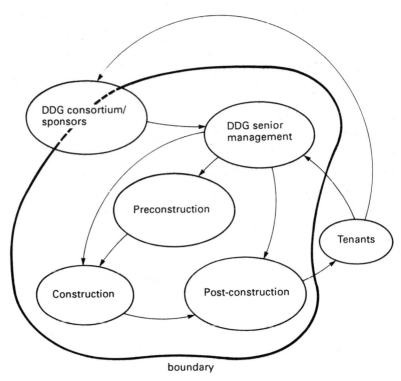

Fig. 14.6 Interaction of variables in the DDG project system.

Category 2: construction
The quality of pre-cast panels and units supplied by prefabricators was not monitored and neither was the quality of workmanship on site. If indeed DDG supplied a Clerk of Works, his or her monitoring was unsatisfactory.

Category 3: structural failures after completion
The most obvious failure relates to the structural failure of pre-cast panels in the post-construction phase. Many concrete panels did not fit together properly and the resulting gaps could not be made watertight. Rubber seals protecting concrete joints linking the panels were sometimes poorly fitted or simply missing which left joints open to rain penetration. Careless packing of mortar between walls and floors allowed water to pour into rooms below thus encouraging mould growth. Patches of cold concrete inside rooms arose from faulty insulation and caused condensation which was aggravated by poor heating and ventilation. Misshapen panels ensured that some window frames fitted badly so that rain seeped in.

Some concrete links suffered load cracks which exposed steel reinforcement and this rusted due to water penetration.

Category 4: management failures after completion
The four chairmen met only once in fourteen years and effectively abandoned DDG tenants. After DDG disbanded in 1968, the four city councils involved took over the management of DDG properties in their respective areas. When tenants and tenants groups such as DDTAG complained over many years, they received little or no redress.

(i) Appropriate models
In addition to the formal system model, apparently relevant models from the repertoire are as follows:

Failure category	Suggested model	Justification
1	control	lack of control action lack of monitoring
	communication	builder's failure to perceive difficulties
2	control	lack of monitoring
3	fault tree	structural failure
	cascade	chain of events
	common mode	similar problems
4	control	controllers absent
	communication	misunderstandings
	stress	GPs' diagnosis
	group behaviour	concerted action by tenants

Stage 2: Comparison

Five comparisons were made between the systems description and the selected models. These comparisons were made as described in Chapter 13, i.e. superimposing aspects of the systems description on the models as shown in Figs 14.7 to 14.11.

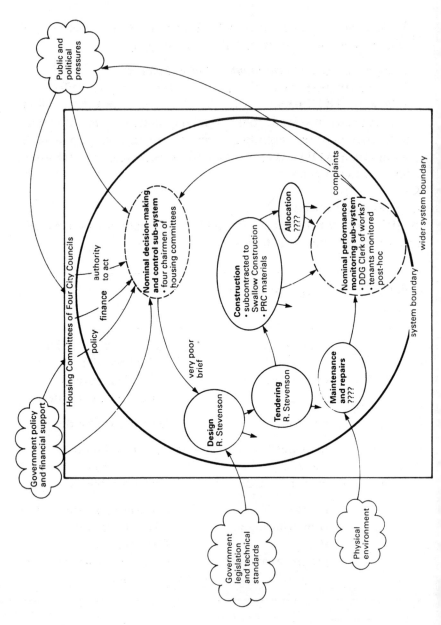

Fig. 14.7 The DDG project system superimposed on the formal system model (second iteration).

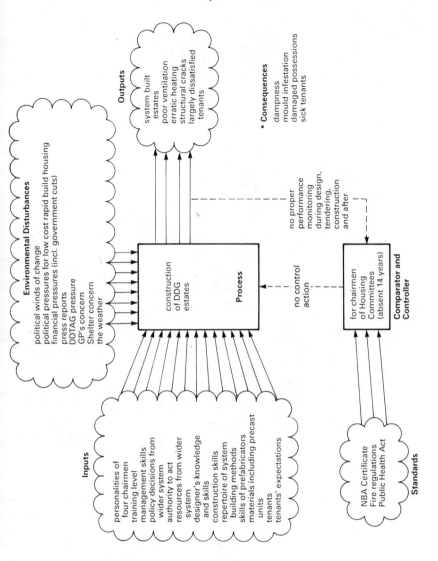

Fig. 14.8 The DDG project system superimposed on the control model (second iteration).

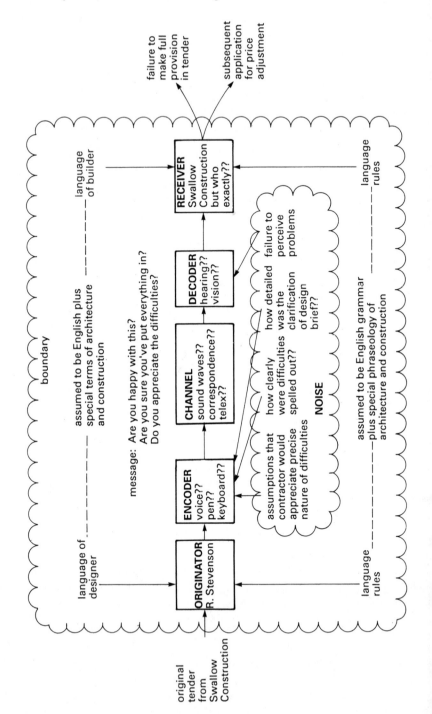

Fig. 14.9 DDG designer's clarification of contractor's tender superimposed on the communication model (second iteration).

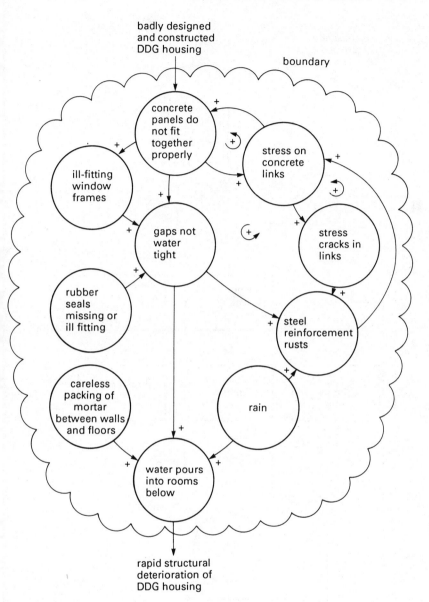

Fig. 14.10 Structural failure of DDG system-built units superimposed on the cascade model (second iteration).

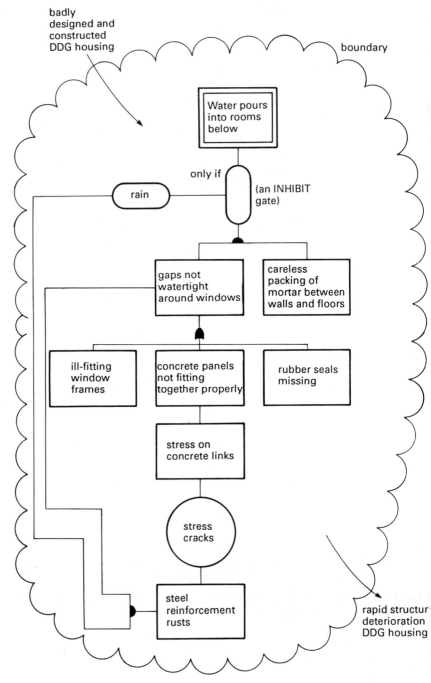

Fig. 14.11 Structural failure of DDG system-built units superimposed on the fault tree model (second iteration).

Stage 3: Meaning of comparisons

Formal system model vs. the DDG Project System

	Model	
	Desirable	Undesirable
No/few discrepancies		
Many discrepancies	X	

There were considerable discrepancies between the DDG Project System and the FSM. For example, the fact that the four chairmen met only once in 14 years suggests that the control sub-system was nominal rather than actual. Their instructions and guidance to operational sub-systems seemed to be laissez-faire and there is no evidence of control beyond a perfunctory briefing of the designer. There also seemed to be no formal performance monitoring sub-system in operation throughout the entire DDG history. Monitoring only arose in an informal way post hoc in the form of tenants' observations and complaints.

Control model vs. the DDG Project System

	Model	
	Desirable	Undesirable
No/few discrepancies		
Many discrepancies	X	

Two major mismatches were (a) no proper monitoring during design, tendering and construction; (b) no control action during design, tendering, construction and the post-construction phase until tenants' complaints became irrepressible. These mismatches are consistent with the FSM interpretation.

Communication model vs. designer's checking of Swallow's tender

	Model	
	Desirable	Undesirable
No/few discrepancies	X	
Many discrepancies		

The data are insufficient to confirm a mismatch. There appears to be a match because the designer asked questions of Swallow Construction and got answers. What is unclear is:

- what was the communication channel?
- the originator's assumptions
- precision and detail of originator's encoding
- how detailed the clarification of the brief was
- who the receiver at Swallow was
- why the receiver failed to perceive difficulties.

Further iteration involving data collection would be needed to pursue this apparent failure.

Cascade model vs. structural failure of panels culminating in water pouring into rooms

	Model	
	Desirable	Undesirable
No/few discrepancies		X
Many discrepancies		

There is a direct match between the system of failures present in the structural deterioration of the DDG units and the cascade model. The particular cascade is more than a simple serial or domino chain in view of the evidence of positive feedback loops, i.e. the more the concrete panels fail to fit together properly the greater the stress on the concrete links; the greater the stress on the concrete links the more the panels do not fit together properly, and so on.

Fault tree vs. structural failure of panels culminating in water pouring into rooms

	Model	
	Desirable	Undesirable
No/few discrepancies		X
Many discrepancies		

As expected from the cascade comparison, there is a direct match. From the fault tree comparison it is harder to appreciate the self-sustaining nature of the failure process. However, the logic of the fault tree does highlight the original, underlying faults and so suggests which ones should be tackled first.

Stage 4: Lessons

The failures of the DDG Project System offer the following lessons:

(1) Any large scale project of this type ought to be under proper management, i.e. competent direction and control. There should be effective monitoring of every phase and aspect of the project.
(2) Communications between designers and building contractors should be seen to be on a formal written basis where important items are concerned. The designer should be obliged to detail explicitly the foreseeable, potential construction difficulties for contractors prior to their tender.
(3) There should be strict quality control checks on prefabricated units received from factories and strict on-site monitoring of the standard of construction.

There is evidence that management, communications and engineering reliability failures were widespread throughout system building projects of the 1960s and 1970s. Although apparently more reliable open system methods are used nowadays, there will always be potential for new build failures unless projects are under sound management with effective communications and engineering reliability procedures.

14.6 Potential for continuation

At this stage, the analyst called a halt owing to time constraints. Limited understanding had been gained of public sector 'new build' housing failures which would be of interest to the Housing Director. Further iterations involving further data collection would enable more careful examination of the DDG Project System, if the client required it.

Beyond a failures approach, the hard systems approach could be used to examine and suggest improvements to engineering reliability of structural components. One aspect of new build failures that became apparent in the wider examination of public housing was the uncertainty about who has actual control over such building projects. Usually, the Housing Department is the client of the Architects Department (see Fig. 2.4, Chapter 2) but the analyst got the feeling that the Architects made most of the decisions. However, the Hulme Conference Report suggests that the public perceive Housing Departments and Housing Committees as being responsible for housing design failures. A soft systems study would be very appropriate for examining organisational relationships, political processes and power distribution between Housing and Architects and how these could be better harnessed to improve the outcomes.

14.7 Selected references

Deck Access Disaster (1985), Report of the Hulme Conference 22 February 1985, City of Manchester Council, May 1985.

Foster J.S. (1983), *Mitchells Structure and Fabric*, Part 1, Batsford Academic and Educational, London.

Marsh P. (1985), Building systems and portable buildings, in *Specifications: Building Methods and Products*, Vol. 1, *Technical and Product Information*, pp. 259–262, ed. D. Martin, The Architectural Press, London.

14.8 The Littlebrook D Hoist Failure – a study of safety management failure

The construction and building industry (civil engineering, house building, building maintenance etc.) has long had the highest incidence of fatal and major injuries per employee of all industries. For example, according to the Health & Safety Executive, from 1981 to 1985 the incidence of such injuries in construction increased steadily from 164.0 to 231.8 per 100 000 employees whereas the incidence for all industries barely altered (60.3 to 63.1). Since the early 1960s, four sets of Construction Regulations have required systematic safety procedures on site. Among these legal requirements are the need to appoint competent persons to oversee safety arrangements and to ensure regular inspection of lifting and other equipment. Many such inspections have to be done either daily or weekly whereas others such as thorough examinations have to be done either six-monthly or every two years. A series of statutory registers recording these inspections and the results have to be maintained on site. In addition to and overlaying these specific requirements, the Health & Safety at Work etc. Act 1974 places a duty on employers to ensure so far as is reasonably practicable the health and safety of all their employees. They also have to draw up and revise as necessary a statement of their health and safety policy and the practical arrangements for carrying it out. In short, employers are expected to manage health and safety in a business-like fashion.

Littlebrook D was a large power station under construction at Dartford, Kent. The CEGB had contracted a major part of the civil engineering work including construction of the cooling water system to John Laing Construction Ltd. They in turn had subcontracted the construction of the shafts and tunnels to Edmund Nuttall Ltd. On 9 January 1978, a hoist operated by Nuttalls failed. The hoist cage fell more than 100 ft (30 m) to the bottom of a 200 ft (60 m) shaft. Four men died and five were seriously injured. The Health & Safety Executive conducted an investigation, the report of which (The Hoist Accident at Littlebrook D Power Station, 9 January 1978) is available from HMSO and booksellers.

Nuttalls were prosecuted and convicted under health and safety legislation. Subsequently, civil actions by those injured or by dependants were also successful. Considerable press publicity was given to what many regarded as a symptomatic failure of the construction industry in particular and employers in general to manage health and safety responsibilities effectively.

14.9 Analysis

Stage 1: Systems description and model selection

Awareness

The analyst was a safety consultant who wanted to gain greater understanding of the Littlebrook D hoist failure and to see if there were any lessons to be learned. In particular, he wanted to demonstrate to clients how such failures occur and how they can be prevented. His own W/a was very much influenced by the safety principle of 'self-regulation', i.e. those who create risks should manage them in a business-like fashion and those who work with such risks should cooperate in this effective management.

The official Health & Safety Executive (HSE) report concluded that the hoist's wire suspension rope broke at a part weakened by corrosion and devoid of lubricant. The deterioration occurred over a relatively short period and was not detected. Analysis of water in the shaft showed that it contained salt and the corrosion was consistent with the rope having been impregnated with salt water. Over the whole of the corroded length of rope at both sides of the fracture, the wire rope core was very corroded and many wires were broken. Tests revealed that at the fracture all the wires had lost over 50% of their tensile strength and the outer wires had lost 80%. In addition, both clamping units of the cage safety mechanism (fall arrestor) were corroded and coated with a hard cement-like deposit weighing an estimated 180 lb ±50 lb. This deposit was visible evidence of the harsh service conditions and its presence should have alerted site staff to have high maintenance standards. At the time of the accident, the hoist cage was carrying nine passengers (one more than the maximum of eight specified).

The hoist manufacturers, ACE Machinery Ltd, specified in their handbook a monthly maintenance procedure including a rope examination. However, 'the exact pattern of routine maintenance could not be established since the contractors kept no record of such work. Site enquiries showed that maintenance work was carried out on a haphazard basis, mainly tending to coincide with the occasions on which repairs to the hoist were required. In the absence of even a simple site record, it was impossible to ascertain whether or not the hoist had been maintained in accordance with the manufacturer's handbook.'

Statutory six-monthly thorough examinations had been delegated to insurance company engineering surveyors (a common and acceptable practice). Such an examination had been due on or before 16 December 1977 but for various reasons the examination did not take place. Since the legal duties lay on Nuttals, they should have either arranged for an examination by another competent person or taken the hoist out of commission until such examination had been done. Up to 9 January 1978, statutory weekly inspections had been carried out by Nuttall's own site staff and at each inspection the entry for the hoist in the register read as 'in good order' or 'in good working order'.

Although Nuttall's safety policy set out general aims, it gave no clear indication about how these were to be achieved. 'Failure to specify detailed procedures was evident throughout the chain of management responsibility and put another burden on the site agent who, in addition to maintaining production rates and overall progress, was expected to ensure that "all machinery and plant ... are maintained in good condition" and to maintain the site's registers, records and reports. Neither the site agent, his deputy (who had been appointed as site safety officer) nor the office manager had received clear guidance, training or allowance of time to undertake this responsibility.'

Although given overall responsibility for the site, the site agent 'clearly needed to be given adequate resources including advice, instruction and appropriately trained staff and suitable support by the organisation'. Assistance on safety matters was intended to be available from a safety coordinator at Nuttall's headquarters who had visited Littlebrook D 13 times over the preceding 18 months. It is not clear whether assistance was requested or what assistance if any was rendered. The failure situation was summarised as a rich picture (Fig. 14.12) and as a spray diagram (Fig. 14.13).

Commitment

The analyst, perceiving that his clients would be in positions comparable with Nuttall's managing director, induced the following commitment:

> 'This accident has come as a terrible shock to the Board. Apart from the tragedy itself, our good name and reputation are at stake. We thought our safety policy which is signed personally by me was a very good one. As MD, I am ultimately responsible for both production and safety in all our operations. I need to find out what went wrong with our safety organisation.' (Nuttall Managing Director).

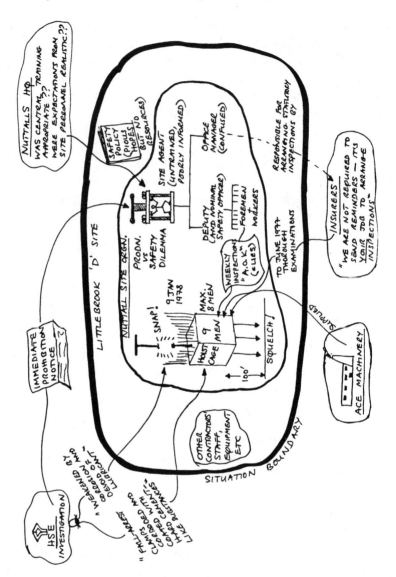

Fig. 14.12 A rich picture of the situation surrounding the hoist accident at Littlebrook D Power Station site (first iteration).

Other potential clients could have been in positions comparable with the managing director of Ace Machinery, the site agent or other key figures but these possibilities were not pursued.

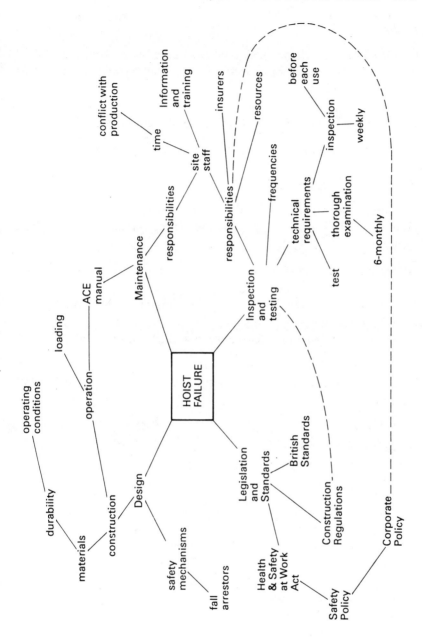

Fig. 14.13 Spray diagram of the situation surrounding the hoist accident at Littlebrook D.

Testing

- Is the potential client's task relevant to the failure situation? (Y)
- Is the commitment (of analyst and potential client) goal-oriented? (Y)
- Is analysis more appropriate than action? (Y)
- Is analysis need-driven rather than just to satisfy curiosity? (Y)

The test results indicate that the situation is worthy of systems analysis.

Separation of systems

One important area in the situation that would be of interest to a potential client is safety vs. production. Three systems relevant to this area are:

A: the Nuttall system for the safe excavation of tunnels at Littlebrook D

B: the Nuttall system for purchasing, operating and maintaining lifting equipment

C: the Nuttall system for formulating and executing safety policy.

In order to save space, only System A is carried here. Typical components and an indication of a trial boundary for System A are as follows:

Within the system	*In the environment*
Managing Director	ACE Machinery
other directors	insurers
safety policy	HSE
resources	CEGB
site agent	John Laing Construction
office manager	HSW Act
deputy site agent	Factories Act
foremen	Construction Regulations
workers	wet, salty conditions
central training dept	
central safety dept	
safety committee	
trades union officials	
safety reps	
plant and equipment	
purchasing	
statutory registers	

System A is a key system because Nuttall's managing director was ultimately responsible for all safety matters in the company. He owns the failure problem. Since there appeared to be a discrepancy between the resources provided for production and those for execution of the safety policy, this system is highly relevant to understanding the failure. Systems B and C were also judged to be key systems.

Selecting a Key System

System A was selected as the first of the three for detailed examination since it relates directly to the hoist accident at Littlebrook D whereas systems B and C are of more general concern.

Detailed Analysis of System A

(a) Components
The resolution of the trial components was adjusted and summarised as a system map (Fig. 14.14).

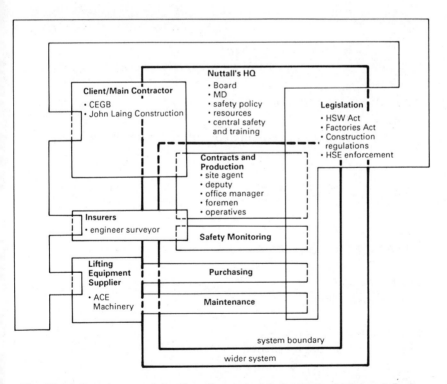

Fig. 14.14 System map of the Nuttall system relating to the Littlebrook D site (third iteration).

(b) Inputs
management skills, knowledge and training
engineering skills
workforce skills, knowledge and training
world-view of Board
world-view of line managers
trades union influence
ACE hoist
client specification.

(c) Outputs
on-shore and off-shore shafts
cooling water tunnels (partly excavated)
ineffective and under-resourced safety policy
poorly trained/informed workforce
poor maintenance
damaged hoist
five fatalities
four seriously injured.

(d) Environment (resolution adjusted)
legislation
client and main contractor
ACE Machinery Ltd
insurers
physical environment (wet, salty conditions, grout deposit etc.).

(e) Structure (resolution adjusted)
Apparent sub-systems were:

board
contracts
purchasing
finance
central training
central safety dept.

(f) Variables (resolution adjusted)
Nuttall's senior management
safety policy
safety resources
behaviour of site personnel

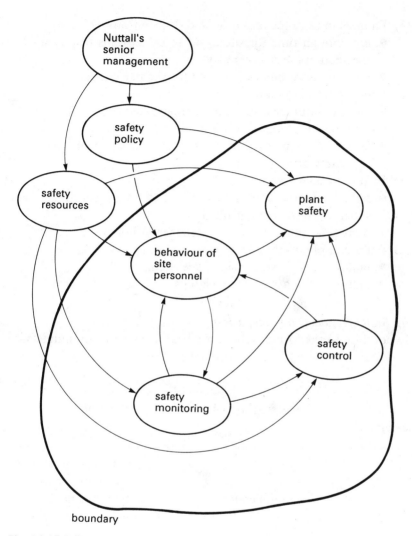

boundary

Fig. 14.15 Influences on the variables of the Nuttall system at Littlebrook D (second iteration).

plant safety
safety monitoring
safety control.

(g) Relationships between variables
Figure 14.15 shows the relationship between variables.

(h) Apparently significant failures
Four failure categories were identified:

(1) Failures in resource provision and distribution
- not enough time for site agent to exercise his responsibilities specified in the safety policy
- no training resources provided for site personnel.

(2) Failures in safety monitoring
- weekly statutory inspections were perfunctory
- six-monthly statutory inspection omitted
- failure of company safety coordinator to adequately monitor site safety arrangements.

(3) Failure in safety control
- failure of company safety coordinator to coordinate and supervise safety arrangements
- no regular preventative maintenance of hoists.

(4) Failure in plant safety
- failure of single suspension rope
- failure of fall-arrest safety clamps.

(i) Suitable models for comparison
In addition to the formal system model, the control, communication and fault tree models are appropriate.

Failure category	Suggested model	Justification
1	control	lack of monitoring and control by site agent
	communication	safety policy not acted upon
2	control	safety inspections casual arrangements not monitored
3	control	no preventative maintenance arrangements not monitored and controlled
4	fault tree	mechanical failures

Stage 2: Comparison

Relevant aspects of the system description were superimposed on the formal system model and the control, communication and fault tree models as shown in Figs 14.16 to 14.19.

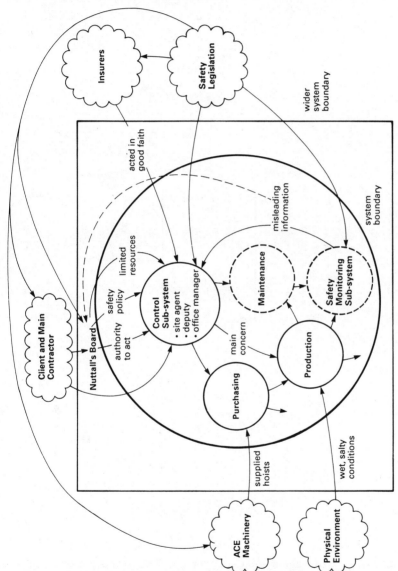

Fig. 14.16 The Nuttall system at Littlebrook D superimposed on the formal system model (second iteration).

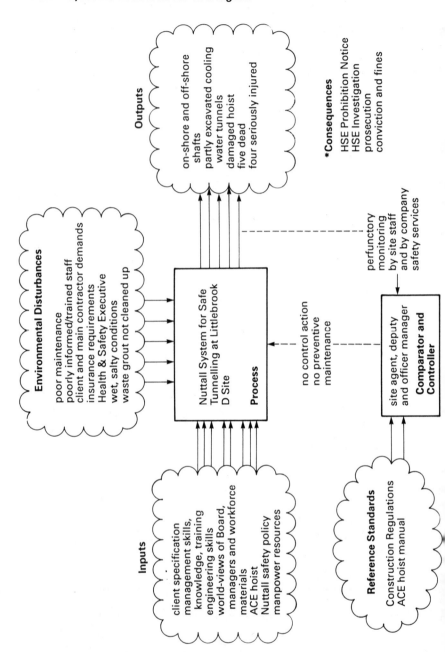

Fig. 14.17 The Nuttall system at Littlebrook D superimposed on the control model (second iteration).

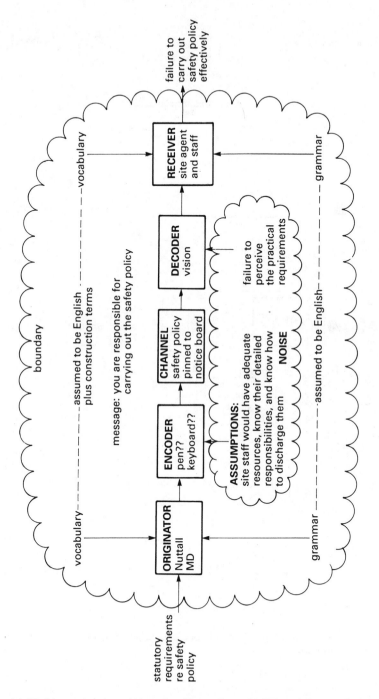

Fig. 14.18 Dissemination of Nuttall safety policy at Littlebrook D superimposed on the communication model (second iteration).

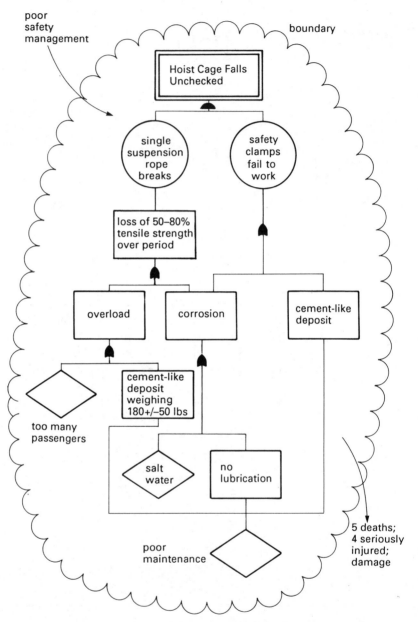

Fig. 14.19 The hoist failure at Littlebrook D superimposed on the fault tree model (second iteration).

Stage 3: Meaning of comparisons

Formal system model vs. the Nuttall system at Littlebrook D

	Model	
	Desirable	Undesirable
No/few discrepancies		
Many discrepancies	X	

Discrepancies include:

(a) from the wider system, an incomplete safety policy and limited resources
(b) control sub-system ineffective owing to overburdened site agent
(c) the maintenance sub-system seemed casual and ineffective (e.g. the cement-like grouting deposit that built up, the corrosion, lack of lubrication)
(d) performance monitoring sub-system seemed casual and ineffective.

Control model vs. the Nuttall system at Littlebrook D

	Model	
	Desirable	Undesirable
No/few discrepancies		
Many discrepancies	X	

Two major mismatches were:

(a) no proper monitoring of safety
(b) no control action to restore proper safety monitoring or to ensure preventative maintenance.

This comparison confirms the interpretation from the FSM comparison.

Communication model vs. dissemination of Nuttall safety policy

	Model	
	Desirable	Undesirable
No/few discrepancies	X	
Many discrepancies		

The information available is insufficient to confirm a mismatch. On the one hand, there appears to be a match because the receivers read the policy documents and appeared to understand the message sent. However, the originator and receivers appeared to have different understandings of what the message meant. More data would have to be collected to examine this aspect further.

Fault tree model vs. faults culminating in unchecked fall of hoist cage

	Model	
	Desirable	Undesirable
No/few discrepancies		X
Many discrepancies		

There is a direct match: a hierarchy of faults culminating in the unchecked fall of the hoist cage.

Stage 4: Lessons

The failures examined offer the following lessons:

(1) Any large scale excavation work ought to be under competent direction and control. There should be effective safety monitoring of every phase and aspect of the work.
(2) Communications of safety policy and directives from senior management (wider system) should take account of the resources of manpower, time, information, instruction, training and supervi-

sion needed to execute those policies and directives effectively on site.

(3) Plant and equipment whose failure would foreseeably result in death or injury should be inspected (i.e. monitored) and should undergo thorough preventative maintenance (i.e. control) at prescribed regular intervals by competent persons. Proper records should be kept and entries should be subject to regular independent cross-checking against the equipment concerned by senior personnel (i.e. monitoring). These monitoring and control activities should be to at least the standard required by law.

14.10 Potential continuation of case study

Systems B and C could be examined in a similar fashion. The focus of attention would then shift to corporate level rather than Littlebrook D. Since weaknesses in management at corporate level emerged in the lessons from system A, examination of systems B and C should provide interesting comparisons. Beyond the study of failures, attitudes towards the 'safety vs production' problem could form the basis of a soft systems study.

14.11 Selected references

The Hoist Accident at Littlebrook D Power Station, 9 January 1978, Health & Safety Executive, HMSO.

Fife I. and Machin E.A. (1982), *Redgrave's Health and Safety in Factories*, second edition, Butterworths, London.

Pinder A. (1983), Safety on construction sites, in J. Ridley (ed.), *Safety at Work*, pp. 630–655, Butterworths, London.

14.12 Summary

The two case studies in this chapter have demonstrated applications of systems failures analysis. Two points emerge from these studies. First, whereas both the cases have been studied by others using non-systems approaches, the framework of SFA provides a disciplined way of gaining understanding from failures and avoids mental 'leapfrogging' in the search for a cause. Second, whereas other methods may be good for certain tasks, they tend to focus on particular, narrow aspects of a failure rather than on underlying causes that interact. SFA integrates model comparisons so that organisational matters are examined on a par with technical ones. With SFM, 'technical faults' are seen to be largely a *consequence* of organisational failures whereas other methods often relegate organisational failures to the category of 'contributory factors'.

Chapter 15
How to Pick-and-Mix

15.1 Introduction

So far, we have described the use of systems methods as if problems and situations present themselves in such a way that the choice of method is clear and obvious. For example, where conflict and unease are apparent, soft systems analysis would be appropriate. If the presenting problem is structured and there is a large measure of agreement about it, then hard systems analysis should be used, and so on. Unfortunately, the analyst is rarely faced with situations where the choice of method is clear cut.

Where the choice is unclear or ambiguous, how should the analyst proceed? Is it possible to apply more than one method during a single study? How would you know which method to use first? Can the analyst safely switch methods in the middle of a study? This chapter addresses such questions.

15.2 Assessing your task

As we have indicated throughout preceding chapters, you as analyst will be concerned not only with world-views in the problem setting but also with your own world-view. Your approach to problems and to applying systems methods (or any other) will be influenced by your W/a. Look back, for example, at the introductory paragraphs to the Daleside case study in Chapter 14 to see how the analyst felt that his W/a had affected his whole stance to public housing. Another analyst would probably have approached the same task from a different angle and with different biases.

In deciding which method to use, there is no requirement that you choose one rather than another. As discussed in Chapters 16 and 17, individuals tend to favour one method rather than another. This is all right provided that natural bias is not taken to such an extreme that you use *only* one method for *all* cases. Your own inclinations and initial assessment of the situation are likely to swing your decision towards one of the three method-types. Intuitive hunches like this are often no worse than lengthy pondering and have an advantage in that you are almost certain to learn something of value from a quick pass with any method. However, there are some guidelines that can improve your efficiency.

15.3 Selecting a method

You may have noticed how many of the examples and case studies in both the 'hard' and 'soft' categories frequently have also a 'failures' character. Interventions in natural systems, such as the Dayak case, and in socio-technical systems such as the Bali case, often produce counter-intuitive outcomes (= failures). Soft problems such as those at Lucrative, Air Traffic Control, and Northbrook could very easily be justified as failures cases. Therefore, if in any doubt when starting a systems study you can never lose anything by trying a failures analysis first. A lot can be learned even if there are no apparently significant failures. You can then decide which if any of the hard and soft methods you are going to apply next.

If there is a large measure of agreement about the nature of the problem (or opportunity) and what is to be achieved, then you may opt for hard systems analysis from the start. Where agreement is low or patchy then soft systems analysis might be a better starting point. Figure 15.1 may be a useful reminder of how the three areas of applied systems (hard, soft and failures) are interrelated.

To recap, you can choose any of the three approaches to start the study but if in doubt use failures analysis first. Having carried out an initial failures study, you can then decide if either or both of the hard and soft methods would be suitable as follow-up studies. However, we do not advocate a slavish approach to using systems methods. Analysts are often up against time pressures that might preclude anything more than a superficial use of even one of the method types. In such cases, it is perfectly acceptable to use systems *ideas* as you are going along rather than trying a step-by-step 'cook-book' approach which will undoubtedly be more time-consuming.

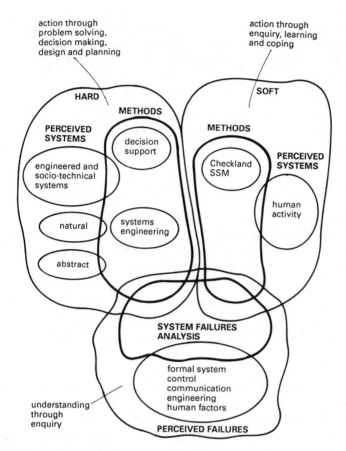

Fig. 15.1 A systems map of applied systems.

15.4 Switching methods

There are occasions when switching methods in mid-stream would be appropriate. Typically, a sensible switching point would be in the early stages of the hard or soft approaches when sufficient description and analysis has been done to indicate that another method would be better. For example, if you start a hard study and discover at systems description stage that failures are manifest, then it may be prudent to find out more about them first so that counter-intuitive outcomes from the hard study may be avoided. Thus, a switch to SFA would be appropriate. Similarly, if you find it impossible in hard systems analysis to clearly identify objectives and constraints it suggests that you do not know enough about the systemic setting; switch to failures analysis to gain understanding. However, it may be that uncertainties and differences of opinion arise about objectives in which case switch to SSM.

If you start with a soft study, your rich picture analysis may throw up apparently significant failures. It would probably be sensible at this stage to switch to SFA to gain greater understanding of the problem situation before returning to SSM. Nevertheless, you may find that when you have identified relevant systems there is a large measure of agreement about them, in which case switch to hard systems analysis. After the debate stage of SSM, it is often useful to conduct a hard study in order to translate the 'whats' of the conceptual model(s) agreed upon into 'hows'. This may engender confidence in the implementation stage.

Of course, the problem situation may not remain stable throughout your period of study. If significant changes occur, this would be another indication for switching methods as you see fit.

Activity

You could, of course, select any of the cases in earlier chapters and try method-switching. We have selected one that you might find interesting. Read the Northbrook case study in Chapter 12. Identify a suitable point to switch to a failures study and carry it out. Assess the understanding gained and consider how it might have affected our soft study of Northbrook.

15.5 Summary

Picking and mixing systems methods is something of an art. There are no rights and wrongs about it but analysts who learn to do it skillfully tend to achieve more effective results than do those who simply resort out of habit to the one method they know best. If in doubt, use failures analysis to start. Applying different methods in series can be very fruitful but sometimes you may have to switch methods in mid-stream. You can do this safely if you follow our guidance above. If time is short, use systems ideas and only as much of the formal methods as you think justifiable to assist you in your task.

PART 3

An Academic Look at Systems Thinking

Chapter 16
The Systems Ideology

16.1 Introduction

As we stated in the Introduction, this is a practical guide to the use of systems methods. The first two parts of this book have concentrated on how to use systems ideas. You have been expected to accept without question that systems methods are available to help with problem solving rather like a set of spanners is available to tackle nuts-and-bolts – an uncontroversial tool-kit. Such an illusion can be convenient. Many people do use systems analytical methods in a taken-for-granted way. However, you need to appreciate that using systems ideas and techniques carries with it a whole set of assumptions that can affect the outcome. In this chapter, we outline and address some of the assumptions of and issues surrounding applied systems. In a relatively short chapter, we cannot do justice to the full range of academic questions and debates and recommend further reading as indicated. We concentrate in this and the following chapter on human activity systems as these tend to be the most complex and the most challenging for applied systems. For those wishing to keep up-to-date with academic thinking in applied systems, we suggest that you read periodicals such as *Systems Practice* and the *Journal of Applied Systems Analysis*.

16.2 Theoretical biases and tensions

The history and development of 'systems' ideas are outlined in a number of references (see for example Burrell and Morgan 1979; Checkland 1981; Cummings 1980; Morgan 1986; Oliga 1988).

The notion of a system is inextricably bound up with an individual's world-view. Checkland (1981) poignantly adds a 'sting' in the tail of his definition of a system: 'a model of a whole entity ... (which may be) applied to human activity. An observer may *choose* to relate this model to real-world activity.' The emphasis is ours to bring attention to a central problem for applied systems, namely the 'as if' problem and its link with world-view. In systems work, 'a system' should refer only to a particular system as a *concept* – not as a 'thing'. However, many concepts frequently do become reified and 'system' is a commonplace example as betrayed in everyday speech; when some individuals refer to the money supply system, for example, they have mental images of concrete components such as the Bank of England, the Stock Exchange and the International Monetary Fund central committee in session. To such individuals, this *is* the monetary supply system. To avoid communication problems arising from different understandings of the term 'system', Checkland (1988) has proposed that it should be replaced by the term 'holon'.

Reification (regarding concepts as if they were things) and use of language are only two aspects that are problematical for applied systems concerning human activity. Two others are world-views about sources of human behaviour and about measurement of human behaviour. When systems analysts and others refer, for example, to organisational systems having 'needs' and 'goals', and exhibiting 'behaviour', these are expressions of a particular kind of world-view that attributes characteristics of an individual human being to an entity that includes a number of individuals – *as if* the entity were an individual. Those who advocate measuring effectiveness of human activity systems solely in terms of 'efficiency in use of resources', 'responses to changes in the economic and market environment', 'market share', 'survival of the fittest' and so on betray a particular kind of world-view – *as if* the system were a biological organism attempting to adapt and survive in a harsh physical environment.

The examples above have illustrated some assumptions linked to particular kinds of world-view. More could have been cited. Analogies and metaphors are, of course, frequently useful provided that their limitations are known and understood. The practical problems that arise from ignoring the fact that 'as if' does not equal 'is' can occur on many planes. A common example is to accept as given that 'the organisation chart' is the organisation. All conceptual models are reductionist. Highly reductionist models such as organisation charts and 'black boxes' convey certain coarse-grained information about structures and processes but little else. Less reductionist models provide fine-grained detail but may overwhelm because everything may seem to be related to everything else.

World-view as a perceptual window biases the individual's perception of systems in general and any particular system. The following sections try to structure these biases.

16.3 World-views about the nature of social reality

A number of reference frameworks have been proposed. The thesis of Burrell and Morgan (1979), for example, is that all theories of organisation are based upon a philosophy of science and a theory of society. They argue that social science may be conceptualised in terms of four sets of assumptions or dimensions, namely:

(1) ontology | the essence of the phenomenon; the degree of objectivity; is it a real object or a concept?

(2) epistemology | the language of description, definition and form; the use of metaphors such as 'system'

(3) human nature | assumptions about the source of human behaviour; is it determined by heredity, culture, and antecedent experience, or by conscious, voluntary, self-created choice?

(4) methodology | the amenability of the variables being measured to particular kinds of instrument; which measures and which methods of data collection are suitable for finding out which information about human activity systems? is human behaviour a series of unique, random events or does it follow a *measurable*, predictable pattern?

Burrell and Morgan (1979) argue that these four dimensions make up the single subjective-objective dimension of world-view, i.e. at the subjective extreme an individual interprets and describes the world in a metaphorical rather than a literal way, is self-determined and measures success in qualitative terms. Conversely, at the objective extreme, an individual interprets and describes the world in a literal, concrete way, sees the role of the individual as serving 'system' interests (e.g. 'corporate man'), and measures success in hard, quantitative terms. A continuum exists between the subjective-objective extremes.

Burrell and Morgan then propose another dimension to cover the range of world-views about social order – at one extreme whether the status quo should be maintained or at the other whether radical change is needed. Burrell and Morgan superimpose this regulation-radical change dimension to cut across the subjective-objective dimension to produce four quadrants that more precisely locate particular world-views.

The interpretive world-view fits in the subjective/regulation quadrant and is consistent with soft systems analysis and other quasi-ethnographic and phenomenological studies. The structural/functionalist world-view fits in the objective/regulation quadrant and encompasses scientific management, operational research, human relations models, and hard systems analysis. Figure 16.1 represents the fundamental differences between the interpretive and structural/functionalist world-views.

The other two quadrants on the radical side in Burrell and Morgan's paradigm are radical humanist and radical structuralist. Advocates of both regard existing social structures as fundamentally unfair. Radical humanists seek to change society by freeing the individual from all forms of oppression whereas radical structuralists seek to transfer power from those groups deemed to possess it to those groups deemed to deserve it (i.e. class-based revolution). See for example Allen (1982) and Clegg (1980). Figure 16.2 represents the two radical world-view sets according to Burrell and Morgan.

The essence of Burrell and Morgan's argument is that the four world-view types cannot be reconciled, i.e. it is not possible, for example, to have a partly interpretive and partly functionalist view of society. This kind of argument is challenged by the theoretical pluralism of Astley and Van de Ven (1983), Donaldson (1985) and Waring (1987) who argue that differing world-views and analysis of the same phenomenon can all be valid and that an individual may simultaneously adopt differing world-views. Waring (1987) argues that an individual's world-view is variable but nonetheless has a prevailing tendency or preference. For example, it is possible for an individual to have a non-exclusive interpretive bias stemming from a world-view that prefers order while recognising that society is essentially unfair and conflictual (i.e. radical nuances). An individual can hold largely subjective beliefs yet recognise that both individual and group

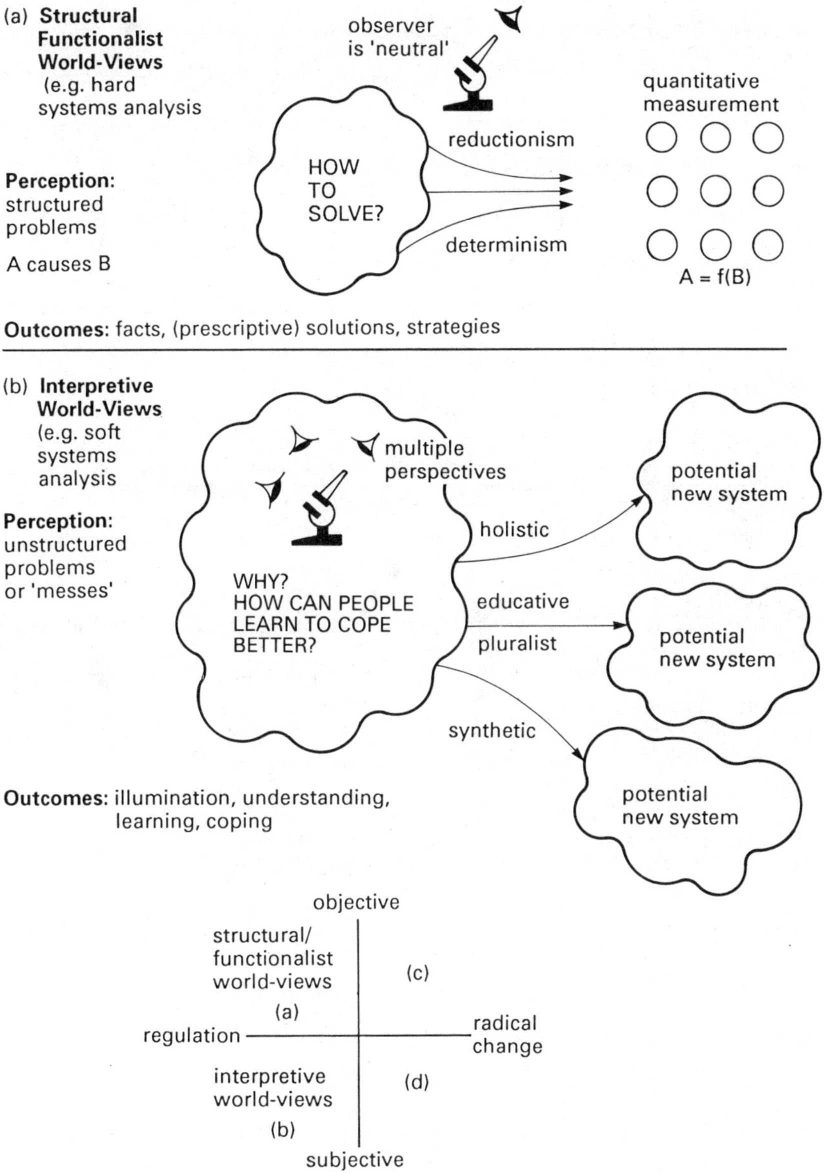

(a) **Structural Functionalist World-Views** (e.g. hard systems analysis

Perception: structured problems

A causes B

observer is 'neutral'

quantitative measurement

HOW TO SOLVE?

reductionism

determinism

$A = f(B)$

Outcomes: facts, (prescriptive) solutions, strategies

(b) **Interpretive World-Views** (e.g. soft systems analysis

Perception: unstructured problems or 'messes'

multiple perspectives

holistic

educative

pluralist

synthetic

WHY? HOW CAN PEOPLE LEARN TO COPE BETTER?

potential new system

potential new system

potential new system

Outcomes: illumination, understanding, learning, coping

objective

structural/ functionalist world-views

(c)

(a)

regulation ——————— radical change

interpretive world-views

(d)

(b)

subjective

Fig. 16.1 Comparison of interpretive and structural/functionalist world-views.

(c) **Radical Structuralist World-Views**

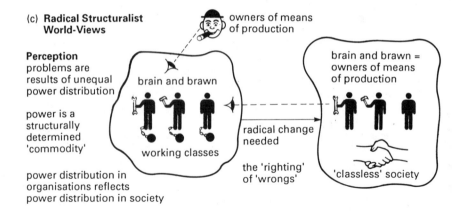

Perception
problems are results of unequal power distribution

power is a structurally determined 'commodity'

power distribution in organisations reflects power distribution in society

Outcomes: class struggle, problems solved by radical socio-economic restructuring

(d) **Radical Humanist World-Views**

Perception:
alienation

the individual is trapped by society

the individual assents unwittingly to own social domination

powerlessness

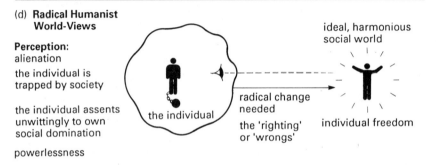

Outcomes: individual action, problems solved by individual resistance to existing socio-economic establishment

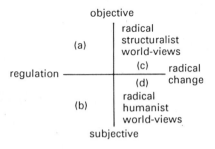

Fig. 16.2 Comparison of radical structuralist and radical humanist world-views.

behaviour have pre-determined components. In other words, an individual with a largely interpretive world-view can still value and apply structural/functionalist arguments. Indeed, the author on the one hand and the two main contributors on the other hand represent two tendencies: the author has an interpretive bias as described, whereas the others have a functionalist bias. These biases affect their preferences for systems techniques but not to the extent that they use their own favoured technique exclusively or even predominantly.

System failures analysis bridges functionalist and interpretive world-view types in that it is essentially a phemonenological method relying on interpretation of real-world phenomena yet also relying on comparisons with functionalist models such as the formal system.

16.4 World-views about the nature of organisations

How do organisations get their characteristics – to be what they are and do what they do? Some authors use a biological system model in which an organisation is likened to an organism competing with others in its environment for limited resources. Natural selection and macro level views (see for example Astley and Van de Ven 1983) have their uses but as exclusive viewing instruments for individual organisations they are too coarse. All organisations will appear undifferentiated because that is what population ecology models intend. Astley and Van de Ven (1983) discuss their paradigm in terms of tensions:

structural forms	vs.	personnel action
part	vs.	whole

These tensions could be relabelled respectively:

structure	vs.	processes
component	vs.	system

without materially affecting the essence of their meaning.

Astley and Van de Ven recognise the problems of theoretical reductionism. The interesting 'fine structure' of observable patterns is lost in attempts to represent them in theoretical models. Echoing Mangham (1979), they note that as a result of training, socialisation and cognitive limits, theorists tend to reduce observed complexities to unidimensional causal models among a limited set of factors that are viewed in isolation from other variables (e.g. scientific management,

OR and similar aspects of structural/functionalism). Reciprocity between explanatory and dependent variables is frequently ignored yet it is such reciprocity that makes tension and conflict between personal actions and structural forms a pervasive characteristic.

How valid are the various paradigms for organisational behaviour proposed or implied in the literature? It is clear that some authors (see for example, Mangham 1979) recognise the theoretical inadequacies of reductionist approaches to organisational behaviour. However, there is a continuing legacy of what has been termed 'biological functionalism' (Burrell and Morgan 1979; Korman and Vredenburgh 1984). Models for both research and managerial practice still regard individuals as passive, rational beings (either totally purposive like ants or totally purposeful and self-willed) instead of as complex psychological, social and economic beings. Organisations are still largely viewed as technical instruments with official goals, strategies, technologies and adaptation to environmental stimuli instead of having complex, dynamic cultures and interactions (see for example: Basil and Cook 1974; Klein 1985; Littler 1985). Where models acknowledge multiple dimensions, the epistemology still reflects a biological paradigm (see for example, the references to 'natural selection' in Astley and Van de Ven 1983). The weaknesses of biological paradigms are discussed further in Section 16.6.

16.5 World-views about individual and group behaviour

How much of human behaviour is pre-programmed (purposive) and how much is consciously self-willed (purposeful)? Purposeful and purposive behaviour may be regarded as alternative terms for subjective and objective behaviour respectively, although Checkland (1981) applies the term purposeful in a looser way to human activity systems to cover what is *assumed* to be purposeful activity. As discussed below, such an assumption may not be justified.

A large proportion (about 90 per cent) of human behaviour results from pre-conscious processing (Dixon 1981), i.e. it is pre-programmed or purposive. Apart from inherited characteristics, purposive behaviour in an individual also incorporates unconscious experiential features resulting from socialisation, cultural norms and so on. Purposeful behaviour, however, involves creative choice. In other words, an individual is not constrained in action by antecedent behaviour or by instinct. Thus, although behaviour is partly purposive (instinct, experience and social conditioning), an individual may exert a wide variety of choice. An individual can break new ground and can think and act purposefully.

Perception, interpretation and meaning involve a complex interplay between purposeful and purposive behaviour. Self-definition and situational definition are partly purposeful and partly purposive in character, as are social interaction and political processes. Even habitual behaviour of interaction where situations are stable, structured and well-understood are not totally purposive. Equally, negotiation in novel, problematic circumstances is not totally purposive. People do not switch on exclusively purposeful behaviour in one situation and exclusively purposive behaviour in another.

The importance of the above comes to the fore in soft systems analysis in which it has been argued (Checkland and Davies 1986; Davies 1988, 1989) that the Weltanschauung of importance is the *collective* one of the social actors. The term 'organisational culture' is used as an alternative to 'collective world-view'. This approach to world-view raises a number of problems (Waring 1988). What, for example, is 'culture'? There are many definitions (Smircich 1983) but most relate in some way to shared ideals, values and ways of doing things. If world-view is to be defined at the level of group culture or collective consciousness, it relies on the assumption that (a) such a 'thing' has an existence independent of individuals (cf. 'public opinion'), (b) the behaviour of individual actors is subordinated and pre-ordained by group culture, i.e. is wholly purposive. Clearly, this level of world-view on its own is unsatisfactory because individuals do not act solely under the influence of 'collective consciousness' even if one accepts such a term as a metaphor for shared values, ideology, rules of behaviour etc. Individuals may agree or assume that they share such attributes but simultaneously, individuals to varying degrees influence events and sometimes *contrary* to what group culture would predict. It is important therefore to address both group *and* key individual world-views in order to avoid introducing another 'as if' problem. This is not to suggest that world-view should be used as a *psychological* viewing instrument on individuals selected as key figures by the analyst.

16.6 World-views about systems of the real world

The range of applications for systems ideas is broad, covering systems engineering, operational research, physical and biological sciences, social sciences, management and organisation, politics and economics. However, since von Bertalanffy first promulgated the modern concept of system in the 1940s and Parsons developed a General Systems Theory in the early 1950s, the emphasis has been on applying hard systems models across the whole range. Such traditional structural/ functionalist ideas are based on the assumption that all systems behave like biological organisms or even machinery and their behaviour can be predicted and therefore controlled. It has become abundantly clear that hard systems models applied indiscriminately as a prescription for better planning and problem solving have been found wanting. The development in the 1970s of soft systems ideas as a complementary, interpretive approach (Checkland 1981) has enabled systems models to be used more selectively. In the late 1980s, however, the predominance of hard systems is still apparent even for applications where soft systems approaches would be more appropriate.

It is important to note that both hard systems and soft systems ideas and approaches are grounded in the same ideological base. They are not fundamentally different concepts but adaptations of the same concept with its roots in biological models. They both assume that humans are essentially rational, ordered beings and that communication, control and adaptation are inherent activities of all systems. In a sense this is so, if one accepts that the vast majority of individual behaviour is purposive.

Systems methods are methods for enquiry and for planning, for producing an outcome (whether understanding or action) that is beneficial to someone. They are not intended primarily to supply knowledge for its own sake as would be the case for traditional scientific enquiry. This does not mean that systems methods cannot be used to advantage as part of human enquiry, provided that limitations are recognised. For example, organisation behaviour is still quite poorly understood whereas soft systems analysis takes for granted that organisations have 'needs', 'goals', and 'cultures'. The tool is perhaps more powerful than the assumptions on which it is based.

As noted above, systems terminology is frequently misapplied both by specialists and the wider public. At worst, the term 'system' is used without qualification to mean method or procedure, i.e. a systematically organised way of doing something. Checkland's (1981) definition of a system is as good a working definition as any: 'a model of a whole entity; when applied to human activity, the model is characterised fundamentally in terms of hierarchical structure, emergent properties, communication and control. An observer may choose to relate this model to real-world activity. When applied to natural or man-made entities, the crucial characteristic is the emergent properties of the whole'. The proposal of 'holon' (Checkland 1988) to replace the term 'system' in systems work is interesting. It is certainly consistent with the holistic concept, however the term 'system' is now so embedded in language that it may take a long time to replace among the thousands of systems practitioners throughout the world.

The adjective 'systemic' is often used in connection with systems viewed in the Checkland way and 'systematic' in connection with methods or other things that do not meet those criteria. Nevertheless, the term 'systemic' is also used frequently in the literature in an inappropriate way. For example, Astley and Van de Ven (1983) explain that 'the systemic argument begins analysis with the organisation as a whole and locates individual action according to its place and function within the system. The individual is only a component of the system, an irritant that must be controlled so that overall functional integration can be maintained.' While such a view may be consistent with 'hard' systems models and traditional structural/functionalist thinking in organisational theory, in research and in managerial practice, its utility is open to question.

As far as human activity systems are concerned, it is important to note that general, or so called 'open', system models that recognise the dynamic interaction between an organisation and its environment are usually of the *hard* variety. See for example de Neufville and Stafford (1971), Morgan (1986) and various authors in Cummings (1980) who use the open system model. Such models include both those that are mechanistic in character and those that are organismic.

The organic or biological paradigms of writers are betrayed in their language of description. For example, Chin (1985) uses such metaphorical terms as 'target system' and 'scanning and sensing mechanisms'. In reviewing the influence of open-systems models in organisational studies, Morgan (1986) notes the assumption of organisational homeostasis, a biological concept denoting capacity for self-regulation and maintenance of a steady state under the influence of an external environment. Biological homeostasis applied to human beings forms part of General Adaptation Syndrome whereby an individual's homeostatic control mechanisms respond automatically (R) to external stimuli (S). Thus, for example, in a moderate thermal environment, gain of heat is balanced by losses largely through radiation, convection, and conduction and also by perspiration (a passive, imperceptible loss of water through the skin). As the thermal environment strays beyond normal limits and heat loading increases, sweating (an active cooling process) is one of several homeostatic responses that enable the individual to adapt to this change in its environment. Self-maintenance is a metaphor that can be recognised in how organisations behave. In most organisations, things tick along in a habitual, predictable way; one does not usually expect to find frequent radical changes occurring at one's workplace. When things get 'hot' in the environment, however, one anticipates that some kind of appropriate adaptive response will be made to prevent damage to the organisation.

However, the homeostasis metaphor is only useful up to a point. On the dynamics of resistance to change, Klein's thesis (1985) is that a necessary prerequisite of successful organisational change is the mobilisation of forces against it. The analogy here is with a biological system such as a human being mobilising its defensive antibodies against antigens invading from the environment. The argument suggests that social systems also exhibit defensive behaviour 'against ill-considered and overly precipitate innovations'. Klein argues that such stimulus-response (SR) behaviour is inevitable, as indeed it would be if the biological model holds true in all cases. However, it could be argued that defensive behaviour within organisations is not automatic or purposive but is dependent upon the values of individual members. Biological homeostasis is an automatic purposive function; *organisational* homeostasis cannot be assumed to be either automatic or entirely purposive as there is a wide variety of possible responses. Some individuals may adopt a defensive role on behalf of the organisation if a threat to their own value system is perceived. Others sharing those values may support them but so too may others with a variety of other other motives. Thus, a biological S-R system model is not an exclusive model for all human activity nor is it necessarily appropriate in any particular case.

Models that liken organisations to living organisms may be useful in certain contexts and for certain purposes. For example, in situations that are stable, structured and well-understood and where general agreement exists about the need for a specific change and the goals to be achieved, a 'hard' systems model for change may be appropriate. However, general systems models assume that organisations have a life of their own, e.g. have 'goals', 'purposes', 'needs' and so on, and therefore they need to be used not as an ideal-type but in such a way that the 'as if' problem is recognised and does not create counter-intuitive outcomes.

The claims by Basil and Cook (1974) that the use of the general systems model removes the biases of the user may be challenged on two grounds. First, the use of *any* model reflects the user's world-view (i.e. biases). Only those with world-views in the extreme objective regions of structural/functionalism and hard science would consider them-selves to be neutral observers whose choice of viewing instrument does not affect their interpretations. Second, on practical grounds the use of such idealised structural, organic models may raise false expectations for success when applied. Inadequate prescriptive models are unlikely to work in practice. As Mangham (1979) puts it, theoretical models of '*the* effective organisation' as proposed by both management scientists and human relations advocates bear little relation to the cultural scripts experienced by members of a real work organisation. Systems analysts have become to be seen as part of particular managerial thinking that is too much concerned with the efficiency of technical 'instruments'. The implications of such biases in applied systems are discussed further in Chapter 17.

Those such as Mumford (1981, 1983) who describe organisations as 'socio-technical systems' fall into the latter category, for such a label suggests the world-view of the traditional systems analyst. It suggests a fundamentally mechanistic or organic model onto which has been tacked a 'social' component. As Dilthey put it (Kluback and Weinbaum 1957), world-views proceed from our conduct of life, from life experiences in general, from the totality of our psychological existence. In Mangham's terms (1979), engineers are biased by their professional experience and tend to apply mechanical conceptual models to all new problems confronting them; traditional systems analysts tend to apply reified ontological paradigms to all situations requiring analysis. These biases are reflected in their language and in their approach to human activity problems.

16.7 Summary

The systems ideology is inextricably bound up with world-views and in particular assumptions about the nature of social reality, the source(s) of individual and group behaviour, the nature of organisation(s) and the processes of social and scientific enquiry. Traditional hard systems ideas and practice, along with the physical and biological sciences and conventional managerial practice, are consistent with a structuralist/functionalist world-view that is ends-and-means dominated. They can have great practical value but success with hard systems models in some applications may lead to false expectations for success when applied to human activity systems.

The challenge addressed by Checkland and followers is to develop systems methods that are capable of coping with inquiry systems, value systems, 'messes', chaotic systems and organisational change. The methods and techniques described in this book represent only a partial snapshot of the current position.

16.8 References

Allen V.L. (1982), *Social Analysis: a Marxist Critique and Alternative*, The Moor Press, Shipley.

Astley W.G. and Van de Ven A.H. (1983), Central perspectives and debates in organization theory, *Administrative Science Quarterly*, **28**, 245–273.

Basil D. and Cook C. (1974), *The Management of Change*, McGraw-Hill.

Burrell G. and Morgan G. (1979), *Sociological Paradigms and Organisational Analysis*, Heinemann.

Checkland P. (1981), *Systems Thinking, Systems Practice*, John Wiley & Sons.

Checkland P. (1988), The case for 'holon', *Systems Practice*, **1**(3), 235–238.

Checkland P. and Davies L. (1986), The use of the term Weltanschauung in soft systems methodology, *Journal of Applied Systems Analysis*, **13**, 109–116.

Chin R. (1985), The utility of the environments of systems for practitioners, in W.G. Bennis *et al.* (eds), *The Planning of Change in Organizations*, 4th edition, Holt, Rinehart and Winston.

Clegg S. (1980), Power, organization theory, Marx and critique, in S. Clegg and D. Dunkerley (eds), *Critical Issues in Organizations*, pp. 21–40, Routledge.

Cummings T.G. (ed.) (1980), *Systems Theory for Organization Development*, John Wiley & Sons.

Davies L. (1988), Understanding organizational culture: a soft systems perspective, *Systems Practice*, **1**(1), 11–30.

Davies L. (1989), letter to the editor, *Systems Practice*, 2(1), 125–128.

de Neufville R. and Stafford J.H. (1971), *Systems Analysis for Engineers and Managers*, McGraw-Hill.

Dixon N. (1981), *Preconscious Processing*, John Wiley & Sons.

Donaldson L. (1985), *In Defence of Organisation Theory*, Cambridge University Press.

Klein D. (1985), Some notes on the dynamics of resistance to change: the defender's role, in W.G. Bennis *et al.* (eds), *The Planning of Change in Organizations*, 4th edition, Holt, Rinehart and Winston.

Kluback W. and Weinbaum M. (1957), *Dilthey's Philosophy of Existence: Introduction to Weltanschauungslehre*, translation of an essay, Vision Press (out of print).

Korman A.K. and Vredenburgh D.J. (1984), The conceptual, methodological and ethical foundations or organisational behaviour, in M. Gruneberg and T. Wall (eds), *Social Psychology and Organizational Behaviour*, pp. 227–254, John Wiley & Sons.

Littler C. (1985), Taylorism, Fordism and job redesign, in D. Knights *et al.* (eds), *Job Redesign: Critical Perspectives on the Labour Process*, pp. 10–29, Gower.

Mangham I. (1979), *The Politics of Organisational Change*, Associated Business Press.

Morgan G. (1986), *Images of Organization*, Sage Publications.

Mumford E. (1981), *Values, Technology and Work*, Martinus Nijhof.

Mumford E. (1983), Successful systems design, in H.J. Otway and M. Peltu (eds), *New Office Technology: Human and Organizational Aspects*, pp. 68–85. Frances Pinter.

Oliga J.C. (1988), Methodological foundations of systems methodologies, *Systems Practice*, **1**(1), 87–112.

Smircich L. (1983), Concepts of culture and organizational analysis. *Administrative Science Quarterly*, **28**(3), 339–358.

Waring A.E. (1987), *New Technology: Managerial Strategies and Employee Responses*, unpublished literature review, The London Management Centre, Polytechnic of Central London.

Waring A.E. (1988), letter to the editor, *Systems Practice*, **1**(3), 323–324.

Chapter 17
Managing Change in Human Activity Systems

17.1 Introduction

This final chapter should be read in conjunction with Chapter 16 as it draws on discussions developed there. Human activity systems are probably the most complex in view of their propensity for behaving in variable and unpredictable ways. Consultants talk blithely of their proprietary strategies for change in client organisations as if all that were required was a dose of the 'prescription' plus a bit of faith healing. Such approaches have all the hallmarks of structural/functionalist world-views and the implication that change is an antidote to some vague unspecified organisational disease. The metaphor may be appropriate in some cases but in this chapter we challenge some of the assumptions of change models and strategies. We suggest how systems methods may be used not as the prescriptive response to a need for change but as tools complementing others in an open-minded non-prescriptive way.

17.2 World-views about organisational change

Where do changes in organisations come from? Basil and Cook (1974) see the stimulus for change coming from the extra-organisational environment in terms primarily of technology change (technological determinism) and to a certain extent changes in social patterns and structures. This stimulus-response approach is consistent with their biological paradigms. However, it is argued that invention and technological developments per se do not stimulate change. Rather, their perceived implications for and effects on the product market, labour market, costs etc. as expressed in Child (1984) stimulate change. Not all stimulus for change originates in the organisation's environment (contrary to the predictions of biological paradigms). Purposeful and even radical decisions for change may arise within an organisation when no substantive pressure from without is perceived. Indeed, discoveries and unplanned inventions of new product types may occur that have a dramatic effect on the product market in the organisation's environment – a reversal of the S–R paradigm. Such an observation is consistent with strategic choice views which assume that the environment is not a set of intractable constraints and can be 'changed and manipulated through political negotiation to fit the objectives of top management' (Astley and Van de Ven 1983).

For organisational change to begin, at least one person possessing power of some kind (as articulated in Mangham, 1979) has to perceive a need for change in the context of his or her definition of the situation. That person has to perceive some personal benefit to be gained in terms of career, reward or status. At this early stage, the attitudinal and decision-making processes are largely purposeful, although any pre-existing socialisation and world-view will exert an unconscious purposive influence on the individual.

To describe this person as the 'change agent' is to accord him or her too much credit for whatever change transpires. He or she may be no more than an initiator, even though the person may have a larger role in later stages. Beyond the initial perception of the need for change, all else is contingent upon social interaction and political processes (see for example: Huff 1980; Mangham 1979; Pettigrew 1985, 1987; Pfeffer 1981, 1982; Quinn 1980; Watson 1982). These processes are in turn contingent upon the relative power of individuals and the groups to which they belong (especially interest groups). They also depend on the skills of those who engage in micro-political processes and their definitions of self and situation. In political processes, behaviour is both purposeful and purposive; individual self-interest and self-determination superimpose to varying degrees on effects of socialisation, cultural values and norms present in that organisation and in its various groups.

Watson (1969) and Klein (1985) describe a change history in terms of a sequence of phases in the change process which is matched by a sequence of resistant phases. However, such a plausible model only holds in some cases. Homeostatic or self-maintaining behaviour may be a feature of all organisations but homeostasis and resistance to change are not synonymous. Organisational homeostasis involves an element of choice (i.e. purposefulness) whereas Watson and Klein suggest that it is a necessary feature of change (i.e. purposive as in the biological S–R sense). As Blackler and Brown (1987) put it in their assessment of changes such as introduction of information technology (IT), there is no a priori cause-and-effect relationship between IT adoption and any structural and behavioural consequences.

The relative power of actors and the nature of decision-making processes are important features in determining the character of a particular change within an organisation. Pfeffer's analysis (Pfeffer 1981) offers four model-types for the analysis of decision-making processes, namely:

(1) *Rational choice models* which assume
- behaviour reflects purpose or intent
- purpose pre-exists
- purposive choices of consistent actors
- goals and objectives are largely pre-ordained
- search process produces a set of decision-making alternatives
- search only continues until necessary and sufficient conditions have been met.

(2) *Bureaucratic models* which assume
- procedural rationality substituted for search and rational choice
- rules determine choice
- feed-forward or even open-loop control.

(3) *Decision process models* which assume
- choice determines preferences
- goals loosely defined, if at all
- retrospective rationalisation.

(4) *Political models* which assume
- action is the result of political process
- when preferences conflict, relative power of social actors determines outcome
- important factors are:

who are the actors?
what determines each actor's stand?
what determines each actor's relative power?
how does decision process work?
how do various preferences become combined?

These models are largely purposive in character, although the political models (see also Huff 1980; Mangham 1979; Pettigrew 1985, 1987;

Watson 1982) are less so. In terms of systems world-views, the rational choice and bureaucratic models are consistent with 'hard' systems approaches. The political models are consistent with 'soft' systems approaches.

Pfeffer's analytical framework also fits that proposed by Earl and Hopwood (1980), although whereas Pfeffer's is concerned with description of theoretical categories Earl and Hopwood's is also concerned with factors that determine the choice of approach to decision making as in Fig. 17.1.

Agreement over objectives

		High	Low
Certainty about cause and effect	High	decision by computation (*rational choice and bureaucratic models)	Decision by politics (*political models)
	Low	decision by judgement (*rational choice and bureaucratic models)	Decision by inspiration (*decision process models)

Fig. 17.1 A decision-making framework. *Pfeffer's model categories superimposed on Earl and Hopwood's analytical framework.

A further theoretical aspect of organisational change needs to be considered. Astley and Van de Ven (1983) support Gomberg's conflictual model of how organisations change. The essence of Gomberg's argument is that organisational structure necessarily generates pressure for its modification among the organisation's members. The conflictual tension grows until a new, but temporary, structure of relationships is synthesised. The cycle of conflict and synthesis repeats itself. This model may be criticised for the following reasons. First, it implies that 'innovating organisers or entrepreneurial managers' are homogeneous and are the only ones capable of initiating change. Second, the purposive S–R nature of the model requires that all organisations stimulate conflict between structure and personnel action and such an absolute view may not be justified. Third, as conflict is not defined, the model is open to ambiguous interpretation. Fourth, the model implies that the synthesis of new relationships automatically leads to a reduction in tension and conflict before the next phase of mounting tension and conflict and this may not always occur.

Mangham (1979) offers a different emphasis. He argues that organisational change occurs as a result of changes in membership, definitions (by individuals or by groups), goals (or low attainment of goals), and roles. These changes require adjustments, i.e. a series of revisions by the process of negotiated order. However, change depends on the location and strength of power. In contrast to the conflictual model, this view does not hold that all members possess equal power (however defined) or have shared values or motives (see also Pfeffer 1981; Watson 1982).

As Pettigrew (1973) notes, change has no terminal state even though terms such as 'introduction' and 'implementation' suggest some clearly defined time markers. Building on the political/cultural arguments of Mangham (1979), Pfeffer (1982), Watson (1982) and others, and on the logical incrementalist approach of Quinn (1980), Pettigrew (1985, 1987) proposes a simple model for understanding organisational change as shown in Fig. 17.2.

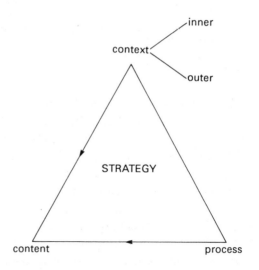

Fig. 17.2 Pettigrew's model of strategy for change.

This elegant model suggests that the content of a change strategy (i.e. what is to change such as technology, manpower, products) arises from the combined effects of inner and outer contexts (i.e. why change is to occur) and processes (i.e. how change comes about such as slow political learning within the organisation). Figure 17.3 attempts to

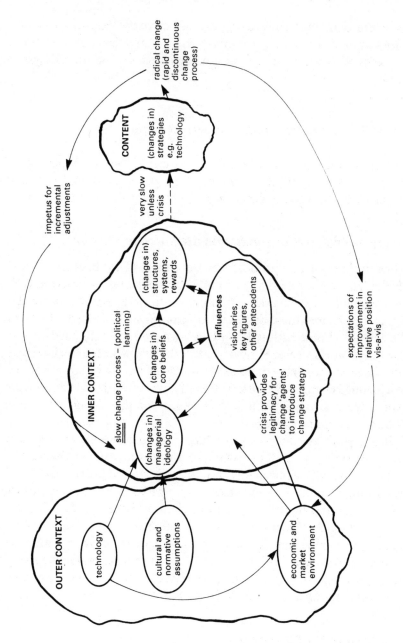

Fig. 17.3 An expanded interpretation of Pettigrew's change model.

represent diagramatically the thrust of Pettigrew's model. Note how although systems ideas are implied, there is avoidance of traditional concepts such as 'environment' and reference to the formal system model.

17.3 Strategies for organisational change

The most striking feature of the literature is that there is no shortage of interpretations of strategy types or of proposed methodologies and strategies for organisational change. Only a selection is outlined here.

Recognising and/or modifying attitudes to change

None of the strategies proposed by Benne and Chin (Benne and Chin 1969; Chin and Benne 1985) is capable of predicting the outcome with any reliability. The location of an individual or group in a particular setting determines the individual's or group's set of attitudes towards the change. These attitudes both inform and are informed by the respective world-views. Similarly, attitudes and world views interplay with heuristic cost-benefit calculations about the likely consequences of a planned change. The distribution of attitudes and their expressions ranges from strong favour and advocacy of the change to strong disapproval of and opposition to it.

It is argued that among many factors influencing the distribution and strength of attitudes of workers towards a planned change the attitudes and behaviour of key figures in the change process will be of prime importance. For example some, perhaps many, individuals value being consulted about proposed changes that may affect them. If they perceive that their concerns and interests are being ignored by managers or other key figures, they may develop an adverse attitude towards the change. In some cases, an adverse attitude focussing initially on a particular change to a particular aspect of their job may generalise into an adverse attitude to the job as a whole.

Change agent intervention

Management consultants rely explicitly or implicitly on their role as change agents in client organisations. Some adopt an interpretive, 'therapist' approach whereas others go for a prescriptive, 'treatment' approach. The latter applied indiscriminately is the most problematic for it is based on an assumption that the organisational 'problem' is clearly identified and well understood and so is amenable to a known 'cure'. A recent example of the role of a change agent in change in a television company is furnished by McLoughlin *et al.* (1985).

The most numerous methodologies are those aimed at the introduction of computers and other forms of information technology into offices. These system design methodologies include the following:

- ETHICS (Effective Technical and Human Implementation of Computer Systems; a socio-technical systems approach developed by Mumford 1981)
- LSDM (Learmonth and Burchett Management Systems Structured Development Method)
- SSADM (Structured Systems and Design Methodology; LSDM developed for the Central Computer and Telecommunications Agency)
- MULTIVIEW (a hard-and-soft approach developed by Wood-Harper and Antill)
- YSDM (Yourdon Structured Method)

Deterministic interventions assume that success is predictable if certain variables are manipulated, e.g. stage-managing technical requirements and resource requirements, changing organisational culture, enabling staff participation and providing staff training (see for example Green 1988). Whereas 'hard' components such as technology and resources may be readily manipulated, 'culture' is almost by definition not amenable to rapid change and especially by outsiders. Even when a newcomer joins at the head of the organisation, it can take years to effect major, lasting change (see for example the summary of Sir John Harvey Jones' efforts at ICI in Pettigrew 1985 and 1987).

A recent development outside the IT field is the application of structured systems analysis and design principles to education and training. Dr Mike Oatey at Southbank Polytechnic has drawn a number of parallels between implementation problems of information systems and those of learning systems. As a result, he has developed a structured analysis and design method that seeks to avoid problems in learning systems.

Participation and related strategies

Various strategies have been proposed for avoiding adverse attitudes towards change in work and for generally smoothing the passage of a change. New technology agreements and participation schemes have enjoyed considerable advocacy (see for example: Blackler and Brown 1987; Mumford 1983; Thompson 1985). However, the rationale of participation is based on a set of assumptions about organisation that are rarely met in practice. True participation requires an equal sharing of decision-making power at all levels of the organisation. In reality, participation is usually a euphemism for consultation or having views taken into account by those who hold real decision-making power. Mangham (1979) refers to such strategies as pseudo-agreement and pseudo-participation. He also notes that, paradoxically, true participation is *less* likely to facilitate change. The barriers to change are represented by the interests of other people and therefore the more power that is shared the more difficult it is to effect change. Hartley (1984) also notes that where true participation is attempted it cannot ensure a balance of expert and referant power and frequently managers and key figures possess more knowledge and information than workers with whom they are supposed to share power. The implication is that participation is an illusory concept and the application of participation strategies in some circumstances may arouse levels of expectations for power sharing that are not met. Adverse attitudes and dissatisfaction may then ensue (see for example Storey 1987).

Planning for change

Strategies are analysed further by Bennis (1969), Barnes (1969), Child (1984), Mangham (1979), Schein (1969) and various authors in Mayon-White (1986). Degree of planning for change feature in all strategies. Mangham's unplanned change is consistent with Pfeffer's decision process model; both are consistent with Blackler and Brown's (1987) Model 0 for IT introduction where 'muddling through' is the norm. Not all managements have strategies in practice. They may *say* that they do and may produce policy documents in evidence (the 'official' history) but in practice it may be 'muddle all the way' with short-term contingencies and expedience dictating the decision-making.

Model 0 strategies accord with the conclusions of Dawson and McLoughlin (1986), McLoughlin *et al.* (1985) and Rose and Jones (1985). They are consistent with political contingencies and the knowledge contingency. Although some authors, notably Mangham (1979), Pettigrew (1985) and Watson (1982) acknowledge that information and skill are required for effective negotiation in social interaction, none mention that technical knowledge may be a pivotal factor. The assumption that managers responsible for introducing new technology, for example, have an accurate and detailed knowledge and understanding of what it can and cannot do may not be justified. Although it may not be necessary for each manager to have expert knowledge of the technical aspects of computers and computing, it is argued that they do need sufficient knowledge in order to frame realistic objectives, to ensure that appropriate specialists and other resources are made available if necessary, and to assess the relative worth of different proposals.

The quality of managers and other key figures is therefore important but does not form a central theme in the literature. It is almost as if such persons are seen as a homogeneous group having similar skills, attributes, motivations etc. This view, and the implication that their managerial behaviour is largely purposive, is no more warranted than it is for 'the workers'.

17.4 Prescriptions and their problems

The problem for most prescriptions for organisational change is that they appear to work unreliably. Sometimes they are deemed successful and sometimes not. Many of the prescriptions put forward at first sight may appear to be loaded with common sense but the fact that they are unreliable suggests that they ignore important variables. At a strategic planning meeting in London in 1988, a guest speaker described his consultancy's strategy for introducing automated production into factories. The approach was very rational, plausible and impressive. At the end of his talk, there were numerous questions from the audience. The interaction went well until a factory manager asked what percentage of his applications had been successful. It transpired that there were still a large number of 'teething troubles' and not a single success could yet be claimed!

This is not an attempt to vilify rational approaches to planning and managing changes. Often, managers and consultants are constrained by lack of time and other resources and are under instruction or contract to deliver results. It would be difficult to justify not using some kind of rational prescription or framework but limitations need to be borne in mind.

One likely problem area is the unwarranted assumption that all the actors involved, and particularly key figures, have a shared understanding of such terms as 'strategy' and 'objective' and of what success will mean. There is evidence (Waring 1988) that among managers there is a wide range of meanings attributed to terms that consultants and specialists define in singular way.

The inherent variability of outcome from political processes may explain why current prescriptions are unreliable, for such processes provide the shifting ground for redefinition over time of the meaning of success of a project (Waring 1988). The context–process–content model (Pettigrew 1987) is particularly relevant.

17.5 How can systems ideas help?

Having warned of the limitations of prescriptive approaches to planning and managing change, we have no intention of offering yet another prescription. However, following from Chapter 15, we can suggest some ways in which systems ideas and techniques can be applied to such an endeavour.

Before a change can be planned effectively, a lot of groundwork needs to be done to establish facts and identify areas of uncertainty. Diagramming techniques such as organisation charts, influence diagrams, causal loop diagrams and so on can be very helpful in marshalling information for your own and other people's benefit. Organisation of objectives, whether as a hierarchy or as a table, will be a prerequisite for a plan. Developed iteratively, such diagrams can be a great help to thinking and communication especially if end-users and others affected by the proposed change are involved in developing a plan.

If change has been proposed because an apparent failure has occurred, then a system failures analysis would be warranted but in any event you could choose to do one or two quick iterations to gain insight into the systemic context of the proposed change. If there are clear suggestions of conflict, unease or anxiety about either the subject of the change, the process of the change or its expected outcome, then soft systems analysis could be used as an enquiry method. You could use the method without embroidery or you could use it as part of larger action research programme along with other data collection and analysis techniques such as participant observation and surveys.

Once objectives for the change are clear and agreed and a shared world-view can confidently be assumed, a feasibility study could be carried out. Formal problem solving or hard systems analysis can be brought into play to select the most promising way to reach the objectives. A wide range of modelling aids could be used, particularly computer simulations, to model and test the essential system's sensitivity to changes in variables. Beyond this stage, design and implementation of the practical system can be achieved by selecting an appropriate systems engineering method from the many available, some of which are mentioned above. Iterations would be required throughout as appropriate.

17.6 Summary and conclusion

In this book, we have provided a practical guide to a wide range of applied systems thinking and practice. Other books and academic courses should be investigated for such needs as indicated in the back of this book. However, we hope that you will have gained some practical insight and benefit from learning how to use systems ideas, whether in the realms of human activity systems, designed systems or whatever.

17.7 References

Astley W.G. and Van de Ven A.H. (1983), Central perspectives and debates in organization theory, *Administrative Science Quarterly*, **28**, 245–273.

Barnes L. (1969), Approaches to change, in W.G. Bennis *et al.* (eds), *The Planning of Change in Organizations*, 2nd edition, Holt, Rinehart and Winston.

Basil D. and Cook C. (1974), *The Management of Change*, McGraw-Hill.

Benne K.D. and Chin R. (1969) General strategies for effecting changes in human systems, in W.G. Bennis *et al.* (eds), *The Planning of Change in Organizations*, 2nd edition, Holt, Rinehart and Winston.

Bennis W.G. (1969), Theory and method in applying behavioural science to planned organizational change, in W.G. Bennis *et al.* (eds), *The Planning of Change in Organizations*, 2nd edition, Holt, Rinehart and Winston.

Blackler F. and Brown C. (1987), Management, organisations and the new technologies, in F. Blackler and D. Oborne (eds), *Information Technology and People: Designing for the Future*, British Psychological Society, Leicester.

Child J. (1984), *Organisation – a Guide to Problems and Practice*, 2nd edition, Harper & Row.

Chin R. and Benne K.D. (1985), General strategies for effecting changes in human systems, in W.G. Bennis *et al.* (eds), 4th edition, *The Planning of Change in Organizations*, Holt, Rinehart and Winston.

Dawson P. and McLoughlin I. (1986), Computer technology and the redefinition of supervision, *Journal of Management Studies*, 23(1), 116–132.

Earl M.J. and Hopwood A.G. (1980), From management information to information management, in H.C. Lucas *et al.* (eds), *The Information Systems Environment*, North Holland Publishing.

Green S. (1988), Strategy, organizational culture and symbolism, *Long Range Planning*, 21(4), 121–129.

Hartley J. (1984), Industrial relations psychology, in M. Gruneberg and T. Wall (eds), *Social Psychology and Organizational Behaviour*, pp. 149–181, John Wiley & Sons.

Huff A.S. (1980), Organisations as political systems: implications for diagnosis, change and stability, in T.G. Cummings (ed), *Systems Theory for Organization Development*, John Wiley & Sons.

Mangham I. (1979), *The Politics of Organisational Change*, Associated Business Press.

Mayon-White B. (ed.), (1986), *Planning and Managing Change*, Harper & Row.

McLoughlin I., Rose H. and Clark J. (1985), Managing the introduction of new technology, *Omega*, 13(4), 251–262.

Pettigrew A. (1973), *The Politics of Organizational Decision-Making*, Tavistock.

Pettigrew A. (1985), *The Awakening Giant: Continuity and Change in ICI*, Blackwell.

Pettigrew A. (1987), Context and action in the transformation of the firm, *Journal of Management Studies*, 24(6), 649–670.

Pfeffer J. (1981), *Power in Organizations*, Pitman.

Pfeffer J. (1982), *Organizations and Organization Theory*, Pitman.

Quinn J.B. (1980), *Strategies for Change: Logical Incrementalism*, Richard D. Irwin.

Rose M. and Jones B. (1985), Managerial strategy and trade union responses in work reorganisation schemes at establishment level, in D. Knights *et al.* (eds), *Job Redesign: Critical Perspectives in the Labour Process*, pp. 81–106. Gower.

Schein E.H. (1969), The mechanism of change, in W.G. Bennis *et al.* (eds), *The Planning of Change in Organizations*, 2nd edition, Holt, Rinehart and Winston.

Storey J. (1987), The management of new office technology: choice, control and social structure in the insurance industry, *Journal of Management Studies*, 24(1), 43–62.

Thompson L. (1985), New office technology: people, work, structure and the process of change, WRU Occasional Paper 34, ACAS Work Research Unit, Department of Employment.

Waring A.E. (1988), *Management of Change, and Information Technology*, unpublished research report, The London Management Centre, Polytechnic of Central London.

Watson G. (1969), Resistance to change, in W.G. Bennis *et al.* (eds), *The Planning of Change in Organizations*, 2nd edition, Holt, Rinehart and Winston.

Watson T.J. (1982), Group ideologies and organisational change, *Journal of Management Studies*, 259.

Explanation of Terms

abstract system: one that is essentially theoretical or symbolic, e.g. a computer programming language

actor: a person perceived to play a role in a human activity system; a social 'actor'.

adaptive control: moderating action taken by a system to maintain its output(s) at a desired level by monitoring output(s), comparing with required level, and counteracting any discrepancy; closed loop control involving feedback signals.

algorithm: any well-defined set of operations for solving a particular problem; may be expressed in words or diagrammatically.

black box model: input–process–output in which process is represented by a notional 'black box'; contents of black box may be ignored.

boundary: a conceptual boundary between the components of a system and the system's environment.

cascade model: a model of a chain reaction; simplest cascade is a 'domino' chain but real-world cascades are usually multi-component and involve multiple causal loops.

CATWOE: mnemonic for the test of adequacy of a Root Definition in the Checkland or soft systems analysis; stands for Customers, Actors, Transformation, Weltanschauung, Owner, Environment.

Checkland method: see soft systems analysis.

client: the person who commissions a system study and to whom the analyst formally reports.

client set: all those with whom a (hard) systems study seeks to gain credibility.

closed system: a system that is not perceived to have any interaction with an environment; real-world systems are rarely closed systems.

communication models: models that depict the encoding, transmission and decoding of information; can be human–human, human––machine or machine–machine.

component: an identifiable part of a system's structure.

conceptual model: a theoretical or abstract model.

control: action taken by a system to maintain its activity or output at a pre-determined level; see adaptive and non-adaptive control.

control models: models that represent the components of control and their relationships.

culture: unwritten and usually unadmitted rules of behaviour, ideologies, habitual responses, language, rituals and 'quirks' that characterise a particular group of people; cultures can be identified at different levels, e.g. nations, societies, organisations, departments, interest groups.

emergence: a collection of properties that characterise a particular system but which could not be attributed to any of the system's components in isolation e.g. customer satisfaction; the result of synergy between components; see synergy, holism and systemic.

engineered system: a system of technical components designed and constructed by man; not necessarily a mechanical system.

engineering reliability: a measure of how successful an engineered system is likely to be in completing its mission; usually measured as a probability in the range 0 to 1.0.

engineering reliability models: models that represent various ways in which the reliability of engineered systems may be affected, e.g. fault trees, cascades.

environment: a conceptual area surrounding a system outside its boundary; components in a system's environment affect the system but whereas components in the system may affect those in the environment they have no control over them.

ETHICS: Effective Technical and Human Implementation of Computer Systems, a socio-technical systems approach developed by Professor Enid Mumford.

fault tree model: an engineering reliability model that enables the possible causes of a fault or failure to be charted as a sequence or hierarchy of preceding events.

feedback: signals from monitoring of outputs from a process are fed back to a comparator which checks for discrepancy between ouput level and desired level; characteristic of closed loop or adaptive control.

feedforward: signals representing a desired level of output are fed to an actuator which adjusts the process so as to produce that level of output; characteristic of open loop or non-adaptive control.

formal problem solving: a formal procedure for solving well-structured and understood problems having limited range of known possible solutions; having defined the problem and considered resources available, the task is essentially a cost-benefit appraisal of the possible solutions; not a systems approach.

formal system model: a model that provides the essential framework of components and processes needed for a system to exist and function, e.g. decision, control, monitoring, communication; applies especially to human activity systems.

functionalist world-view: a world-view that suggests that everything has a pre-ordained role and function; the structures of business and industry and how they operate are taken for granted; 'problems' exist to be 'solved'; consistent with hard systems analysis.

hard system: one that has well-structured components and definable, quantifiable attributes which lend themselves to prediction and control.

hard systems analysis: any systems method that assumes that a problem has to be solved or an unmet need or opportunity fulfilled; analysis involves detailed description of systems and a rigorous evaluation of potential strategies; client-set world-view is relevant throughout.

hierarchy: the concept of relationships in which sets of items are subordinate to others in some way, e.g. a family tree, a set of objectives.

holism: a concept often expressed as 'the whole is greater than the sum of its parts'; encompasses the concepts of emergence, synergy and systemic properties.

holon: an alternative term for system proposed by Professor Peter Checkland in order to avoid problems associated with the wide misuse of the term 'system'; crucially, 'what a holon (system) shall contain is determined by the observer'.

homeostasis: self-maintaining activity consistent with adaptive, closed-loop control; characteristic of biological systems but may be applied to human activity systems with caution.

human activity system: one that involves people apparently carrying out some purposeful activity; the system content is conceptual rather than real-world, i.e. it is determined by the observer; usually soft in character.

human factors models: models that represent particular aspects of human behaviour, e.g. learning, stress, group dynamics.

information system: a system designed to provide users with information on prescribed topics.

interpretive world-view: a world-view that does not regard everything as having a pre-ordained role and function although it does value social order; the structures of business and industry and human activity in general are seen as being socially constructed; many real-world 'problems' require coping and learning strategies rather than 'solutions'; consistent with soft systems analysis.

issue: a topic about which there is (overt or covert) disagreement among social actors; a bone of contention.

iteration: the process of repeating actions; a standard activity in systems analysis in order to develop clarity and understanding or improve design.

JSM: Jackson Structured Method; a hard systems approach to information systems analysis and design.

key figure: a social actor in a human activity system whom the observer considers to exercise a special or important role.

LSDM: Learmonth and Burchett Structured Development Method; see SSADM.

measure of assessment: a means of determining the anticipated results of a number of different proposed options in hard systems analysis and formal problem solving; in hard systems analysis, both quantitative and qualitative measures are needed.

mess: a system of problems that defies resolution simply by trying to solve individual problems; typical of apparently intractable 'wicked' problems of human activity systems; the starting point of soft systems analysis.

MRP-II: Manufacturing Resource Planning; a hard systems approach to production management.

natural system: a system that is not man-made and exists in the natural world, e.g. the weather.

non-adaptive control: control that does not rely on monitoring process output and adapting to any discrepancy between output and desired state; feedforward or open-loop control.

open system: a system that interacts with an environment; an adaptive system.

organisation: (1) the processes of forming, maintaining and dissolving human relationships; (2) a structure of such relationships as in 'the' organisation; (3) a human activity system as in 'the' organisation.

owner: (1) a system owner is the person who has ultimate authority over its existence; (2) a problem owner is the person who has the task of resolving it.

paradigm: a model.

PLOT: Production Logistics Organisation Technique; a hard systems approach to production management developed by the Computer Manufacturing Group.

political processes: processes in human activity systems by which various power factors (e.g. authority, influence, knowledge, information, world-views) affect the course and outcomes of action.

power: an emergent property of social interaction within a particular context; a complex phenomenon (see political processes).

primary task: the essential task of a human activity system, e.g. the primary task of a motor manufacturer is to make motor vehicles.

problem: a source of puzzle, annoyance, frustration or harm to someone; problems do not exist outside people's heads.

process: a system component that changes continuously; 'doing' of some kind.

purposeful: consciously willed behaviour involving creative choice.

purposive: pre-ordained or pre-conscious behaviour resulting from a complex mixture of heredity and previous cultural and other experience.

real world: the world outside the artificial world of the laboratory.

reductionism: a process of reducing complexity to simpler and more manageable components; an inherent part of modelling in hard systems analysis; reductionist models assume that all the fine structure that is lost is insignificant to the task.

reification: regarding something as an object even if it is intangible, e.g. the money supply.

relevant system: in soft systems analysis, a hypothetical system deemed by the analyst to be relevant to the mess and ultimately tested for relevance on the actors.

resolution: the degree of detail covered in a system diagram; the number of components needs to be adjusted to be informative and to avoid swamping a reader.

rich picture: an evocative visual summary of a mess; drawn at the start of soft systems analysis but may also be used in systems failures analysis.

root definition: the precise definition of a relevant system in soft systems analysis.

SASDM: Structured Analysis and Systems Design; a hard systems approach to information systems developed by Tom DeMarco and Ed Yourdon.

socio-technical system: a system conceived as having both engineered or designed components and human activity components, e.g. an information system of which computers form a part.

soft system: one concerned with human activity.

soft systems analysis: a method intended for use where a human activity system exhibits crisis, conflict, uncertainty or unease in relationships among the actors; also called the Checkland method.

SSADM: Structured Systems Analysis and Design Methodology; a hard systems approach to information systems developed for the Central Computer and Telecommunications Agency by Learmonth and Burchett Management Systems.

strategy: an overall plan of action to achieve a desired objective.

stress model: a human factors model that represents how a human being adapts to a variety of stressors.

structure: relatively stable and unchanging components of a system; the 'doers' and the 'done to'.

sub-system: an identifiable component of a system that itself has the characteristics of a system.

symbolic system: an abstract system.

synergy: interaction between system components to produce an output greater than the sum of component outputs; see emergence and holism.

system: a concept of a recognisable whole consisting of a number of parts that interact in an organised way; characterised by outputs, emergent properties, a boundary, an environment and an owner; addition or removal of a component affects both the system and the component.

system description: a structured way of describing a system.

system failure: an apparent shortcoming in a system that causes someone concern or is otherwise detrimental to someone.

systematic: an organised way of doing something.

systemic: to do with a system and implying holism and emergence.

systems engineering: a hard systems approach to the design and construction of technical or engineered systems.

transformation: the essential process of a relevant system in soft systems analysis.

value: a belief or set of beliefs that is untestable, e.g. religious beliefs, political views, moral standpoints.

value system: a number of values that interact.

Weltanschauung: world-view; a complex set of perceptions, attitudes, beliefs, values and motivations that characterise how an individual or group of people interpret the world and their existence; characteristic biases.

wider system: a system outside the system that authorises its existence, sets policy, provides resources etc., e.g. a parent company represents the wider system of a subsidiary.

world-view: see Weltanchauung.

YSM: Yourdon Structured Method; a hard systems approach to information systems developed by Ed Yourdon.

Useful Reading

Systems books and courses

Systems Thinking, Systems Practice by Professor Peter Checkland, published by John Wiley & Sons (1981).
Systems Concepts, Methodologies and Applications by Brian Wilson, published by John Wiley & Sons (1984).
Understanding Systems Failures by Victor Bignell and Joyce Fortune, published by Manchester University Press (1984).
Structured Systems and Design Methodology by Geoff Cutts, published by Paradigm Publishing (Blackwell Scientific Publications) (1987).
Systems, Management and Change: a Graphic Guide by Ruth Carter, John Martin, Bill Mayblin and Michael Munday, published by Harper & Row (1984).
Information Systems Design: the Multiview Approach by Trevor Wood-Harper, Lyn Antill and David Avison, published by Blackwell Scientific Publications (1985).
Systems courses (e.g. T241 and T301) from the Open University.

Other references not listed elsewhere in the book

Booth R.T., Raafat H. and Waring A.E. (1988), *Machinery and Plant Integrity*, Module ST2, Occupational Health & Safety Open Learning, Portsmouth Polytechnic.
Boyle A.J. and Wright C.C. (1984), Accident 'migration' after remedial treatment at accident black spots, *Traffic Engineering and Control*, **25**(5), 260–267.
Churchman C.W. (1985), Perspectives of the systems approach, in W.G. Bennis *et al.* (eds), 4th edition, *The Planning of Change in Organizations*, pp. 253–259, Holt, Rinehart and Winston.

Holling C.S. and Goldberg M.A. (1973), *Managing the Environment*, US Government Printing Office.

Oatey M. and Payne C. (1986), *COBOL from BASIC: a Short Self-instructional Course*, Pitman.

Rollinson J.G. (1971), *Creative Thinking and Brainstorming*, tape/slide presentation by Management Training Ltd, Parker Street, London WC2B 5PT.

Rowe C.J. (1985), Identifying causes of failure: a case study in computerised stock control, *Behaviour and Information Technology*, **4**(1), 63–72.

Waring A.E. (1986), *Office Environment*, Module OTA5, Office Technology Open Learning, South Bank Polytechnic.

Index

real world, xiv, 91, 95, 99, 101, 142, 203, 205, 207, 223, 225, 234, 305, 328
reductionism, 56, 296, 301, 302, 328
redundancy, 123, 124
reference value, 111–13
reification, 296, 307
relationship, 20, 28, 244, 259
relevant system, 40, 91, 95, 99, 201, 202, 218, 229, 328
reliability, 120–23
  block diagram, 122, 123
resolution, 23, 24, 168, 241–3, 256, 278, 328
resources, 102, 199, 205, 237, 244, 246, 252, 280, 301
rich picture, 37, 38, 74, 77–81, 88–91, 99, 199, 207, 216, 217, 223, 229, 241, 255, 273, 274, 328
root definition, 91, 95–7, 99, 100, 201–4, 209, 218, 219, 229, 231, 329
runaways, 114, 117, 129, 133

safety factor, 122, 134
self-enhancing, 30
self-maintaining, 30, 31, 35
self-regulation, 272, 305
sensitivity analysis, 184
signing system, 45, 53
simulation, 10, 31, 49
situation summary, 74, 77
snowball effect, 48, 129
social reality, 297, 307
socio-technical system, 16, 104, 290, 307, 329
soft system, xii, 3, 11, 16, 45, 52, 74, 329
  analysis, 74, 198, 199, 246, 289, 290, 303, 313, 329
  approach, xv, 91–3
  method, 198, 208, 212

software, 52, 86, 156, 161
spidergram, 24, 144
spray diagram, 18, 24–7, 37, 144, 165, 166, 239, 241, 275
S-R paradigm, 305, 306, 311–13
SSADM, 35, 36, 101, 190, 329
Stafford, 61
strategy, 59, 150, 152, 153, 156, 172, 173, 181, 190, 310, 314, 316, 317, 318, 320, 329
stress model, 110, 130, 131, 261, 329
structural component, 37, 78, 270
structural failure, 129, 241, 260, 268, 269
structural/functionalist, 298, 299, 301, 304, 305, 307, 310
structure, 45, 74, 258, 278, 296, 329
structure and relationship diagram, 18, 19, 21
structured problem, 92, 299
sub-system, 9, 24, 41, 51, 93, 102, 199, 204, 243, 244, 258, 329
symbolic system, 45, 49, 329
symptoms, 71, 106, 109, 127, 271
synergy, 9, 329
system, 3–6, 296, 304, 305, 329
  behaviour, 49, 68, 74, 246
  description, 18, 36, 144, 240, 241, 246, 252, 272, 281, 329
  design, 317
  diagram, 47, 74, 79, 199
  efficiency, 5, 45, 51, 307
  instability, 31, 48
  map, 18, 21–3, 42, 43, 65, 67, 144, 168, 243, 256, 257, 277, 291
  noise, 110, 114, 118, 133
  owner, 3, 6, 10, 56, 59, 96, 142, 143, 202, 205, 211, 219, 328
  reliability, 120, 122–4, 134
  structure, 6, 243
systematic, xv, 4, 137, 142, 244, 305, 329